SpringerBriefs in Electrical and Computer Engineering

Signal Processing

Series Editor

C.-C. Jay Kuo, Los Angeles, USA
Woon-Seng Gan, Singapore, Singapore

For further volumes:
http://www.springer.com/series/11560

Jia He · Chang-Su Kim
C.-C. Jay Kuo

Interactive Segmentation Techniques

Algorithms and Performance Evaluation

 Springer

Jia He
Department of Electrical Engineering
University of Southern California
Los Angeles, CA
USA

C.-C. Jay Kuo
Department of Electrical Engineering
University of Southern California
Los Angeles, CA
USA

Chang-Su Kim
School of Electrical Engineering
Korea University
Seoul
Republic of South Korea

ISSN 2196-4076   ISSN 2196-4084   (electronic)
ISBN 978-981-4451-59-8   ISBN 978-981-4451-60-4   (eBook)
DOI 10.1007/978-981-4451-60-4
Springer Singapore Heidelberg New York Dordrecht London

Library of Congress Control Number: 2013945797

Printed on acid-free paper

Springer is part of Springer Science+Business Media (www.springer.com)

Preface

Image segmentation is a key technique in image processing and computer vision, which extracts meaningful objects from an image. It is an essential step before people or computers perform any further processing, such as enhancement, editing, recognition, retrieval and understanding, and its results affect the performance of these applications significantly. According to the requirement of human interactions, image segmentation can be classified into interactive segmentation and automatic segmentation. In this book, we focus on *Interactive Segmentation Techniques*, which have been extensively studied in recent decades. Interactive segmentation emphasizes clear extraction of objects of interest, whose locations are roughly indicated by human interactions based on high level perception. This book will first introduce classic graph-cut segmentation algorithms and then discuss state-of-the-art techniques, including graph matching methods, region merging and label propagation, clustering methods, and segmentation methods based on edge detection. A comparative analysis of these methods will be provided, which will be illustrated using natural but challenging images. Also, extensive performance comparisons will be made. Pros and cons of these interactive segmentation methods will be pointed out, and their applications will be discussed.

Contents

Chapter 1
Introduction

Keywords Interactive image segmentation · Automatic image segmentation · Object extraction · Boundary tracking

Image segmentation, which extracts meaningful partitions from an image, is a critical technique in image processing and computer vision. It finds many applications, including arbitrary object extraction and object boundary tracking, which are basic image processing steps in image editing. Furthermore, there are application-specific image segmentation tasks, such as medical image segmentation, industrial image segmentation for object detection and tracking, and image and video segmentation for surveillance [1–5]. Image segmentation is an essential step in sophisticated visual processing systems, including enhancement, editing, composition, object recognition and tracking, image retrieval, photograph analysis, system controlling and vision understanding. Its results affect the overall performance of these systems significantly [2, 6–8].

To comply with a wide range of application requirements, a substantial amount of research on image segmentation has been conducted to model the segmentation problem, and a large number of methods have been proposed to implement segmentation systems for practical usage. The task of image segmentation is also referred to as object extraction and object contour detection. Its target can be one or multiple particular objects. The characteristics of target objects, such as brightness, color, location, and size, are considered as "objectiveness", which can be obtained automatically based on statistical prior knowledge in an unsupervised segmentation system or be specified by user interaction in an interactive segmentation system. Based on different settings of objectiveness, image segmentation can be classified into two main types: automatic and interactive [9].

Automatic segmentation has been widely used in image/video object detection, multimedia indexing, and retrieval systems, where a quick and coarse region-based segmentation is sufficient [9]. However, in some applications such as medical image segmentation and generic image editing, a user may want more accurate segmentation with an accurate object boundary with all object parts extracted and connected.

J. He et al., *Interactive Segmentation Techniques*,
SpringerBriefs in Signal Processing
DOI: 10.1007/978-981-4451-60-4_1, © The Author(s) 2014

In most cases, it is difficult for a computer to determine the "objectiveness" of the segmentation. In the worst case, even with clearly specified "objectiveness," the contrast and luminance of an image is very low and the desired object has similar colors with background, which may produce weak and ambiguous edges along object boundaries. Under these situations, automatic segmentation may fail to capture user intention and produce meaningful segmentation results.

To impose constraints on the segmentation, interactive segmentation involves user interaction to indicate the "objectiveness" and thus to guide an accurate segmentation. This can generate effective solutions even for challenging segmentation problems. With prior knowledge of objects (such as brightness, color, location, and size) and constraints indicated by user interaction, segmentation algorithms often generate satisfactory results. A variety of statistical techniques has been introduced to identify and describe segments to minimize the bias between different segmentation results. Most interactive segmentation systems provide an iterative procedure to allow users to add control on temporary results until a satisfactory segmentation result is obtained. This application requires the system to process quickly and update the result immediately for further refinement, which in turn demands an acceptable computational complexity of interactive segmentation algorithms.

A classic image model is to treat an image as a graph. One can build a graph based on the relations between pixels, along with prior knowledge of objects. The most commonly used graph model in image segmentation is the Markov random field (MRF), where image segmentation is formulated as an optimization problem that optimizes random variables, which correspond to segmentation labels, indexed by nodes in an image graph. With prior knowledge of objects, the maximum a posteriori (MAP) estimation method offers an efficient solution. Given an input image, this is equivalent to minimizing an energy cost function defined by the segmentation posterior, which can be solved by graph-cut [10, 11], the shortest path [12, 13], random walks [14, 15], etc. Another research activity has targeted at region merging and splitting with emphasis on the completion of object regions. This approach relies on the observation that each object is composed of homogeneous regions while background contains distinct regions from objects. The merging and splitting of regions can be determined by the statistical hypothesis techniques [16–18].

The goal of interactive segmentation is to obtain accurate segmentation results based on user input and control while minimizing interaction effort and time as much as possible [19, 20]. To meet this goal, researchers have proposed various solutions and their improvements [18, 21–24]. Their research has focused on algorithmic efficiency and satisfactory user interaction experience. Some algorithms have been developed as practical segmentation tools. Examples include the Magnetic Lasso Tool, the Magic Wand Tool, and the Quick Select Tool in the Adobe Photoshop [25], and the Intelligent Scissors [26] and the Foreground Select Tool [27, 28] in another imaging program GIMP [29].

Each image segmentation method has its pros and cons on different tasks. Performance evaluations have been conducted on interactive segmentation methods, including segmentation accuracy, running time, user interaction experience, and memory requirement [2, 9, 22, 24, 30, 31]. In this book, we discuss the strengths and weaknesses of several representative methods in practical applications so as to provide a guidance to users. Users should be able to select proper methods for their applications and offer simple yet sufficient input signals to the segmentation system to achieve the segmentation task. Furthermore, discussion on drawbacks of these methods may offer possible ways to improve these techniques.

We are aware of several existing survey papers on interactive image segmentation techniques [6, 7, 32, 33]. However, they do not cover the state-of-the-art techniques developed in the last decade. We describe both classic segmentation methods as well as recently developed methods in this book. This book provides a comprehensive updated survey on this fast growing topic and offers thorough performance analysis. Therefore, it can equip readers with modern interactive segmentation techniques quickly and thoroughly.

The remainder of this book is organized as follows. In Chap. 2, we give an overview of interactive image segmentation systems, and classify them into several types. In Chap. 3, we begin with the classic graph-cut algorithms and then introduce several state-of-the-art techniques, including graph matching, region merging and label propagation, clustering methods, and segmentation based on edge detection. In Chap. 4, we conduct a comparative study on various methods with performance evaluation. Some test examples are selected from natural images in the database [34] and Flickr images (http://www.flickr.com). Pros and cons of different interactive segmentation methods are pointed out, and their applications are discussed. Finally, concluding remarks on interactive image segmentation techniques and future research topics are given in Chap. 5.

References

1. Bai X, Sapiro G (2007) A geodesic framework for fast interactive image and video segmentation and matting. In: IEEE 11th international conference on computer vision, ICCV 2007, IEEE, pp. 1–8
2. Grady L, Sun Y, Williams J (2006) Three interactive graph-based segmentation methods applied to cardiovascular imaging. In: Paragios N, Chen Y, Faugeras O (eds) Handbook of Mathematical Models in Computer Vision. Springer, pp. 453–469
3. Ruwwe C, Zölzer U (2006) Graycut-object segmentation in ir-images. In: Bebis G, Boyle R, Parvin B, Koracin D, Remagnino P, Nefian AV, Gopi M, Pascucci V, Zara J, Molineros J, Theisel H, Malzbender T (eds) Proceedings of Second International Symposium on Advances in Visual Computing, ISVC 2006, Nov 6–8, vol 4291. Springer, pp 702–711, ISBN: 3-540-48628-3, http://researchr.org/publication/RuwweZ06, doi:10.1007/11919476_70
4. Steger S, Sakas G (2012) Fist: fast interactive segmentation of tumors. Abdominal Imaging. Comput Clin Appl 7029:125–132
5. Sommer C, Straehle C, Koethe U, Hamprecht FA (2011) ilastik: interactive learning and segmentation toolkit. In: 8th IEEE international symposium on biomedical imaging (ISBI 2011)
6. Ikonomakis N, Plataniotis K, Venetsanopoulos A (2000) Color image segmentation for multimedia applications. J Intel Robot Syst 28(1):5–20
7. Luccheseyz L, Mitray S (2001) Color image segmentation: A state-of-the-art survey. Proc Indian Natl Sci Acad (INSA-A) 67(2):207–221
8. Pratt W (2007) Digital image processing: PIKS scientific inside. Wiley-Interscience publication. Wiley, New York

9. McGuinness K, O'Connor N (2010) A comparative evaluation of interactive segmentation algorithms. Pattern Recogn 43(2):434–444

10. Boykov Y, Jolly M (2001) Interactive graph cuts for optimal boundary and region segmentation of objects in nd images. In: Eighth IEEE international conference on computer vision, 2001. ICCV 2001, IEEE, vol 1, pp. 105–112

11. Boykov Y, Veksler O (2006) Graph cuts in vision and graphics: theories and applications. In: Handbook of Mathematical Models in Computer Vision pp 79–96

12. Mortensen E, Barrett W (1998) Interactive segmentation with intelligent scissors. Graph Models Image Proces 60(5):349–384

13. Mortensen E, Morse B, Barrett W, Udupa J (1992) Adaptive boundary detection using 'live-wire' two-dimensional dynamic programming. In: Computers in Cardiology 1992. Proceedings, IEEE, pp. 635–638

14. Grady L (2006) Random walks for image segmentation. IEEE Trans Pattern Anal Mach Intel 28(11):1768–1783

15. Kim T, Lee K, Lee S (2008) Generative image segmentation using random walks with restart. Comput Vision-ECCV 2008:264–275

16. Adams R, Bischof L (1994) Seeded region growing. IEEE Trans Pattern Anal Mach Intel 16(6):641–647

17. Mehnert A, Jackway P (1997) An improved seeded region growing algorithm. Pattern Recogn Lett 18(10):1065–1071

18. Ning J, Zhang L, Zhang D, Wu C (2010) Interactive image segmentation by maximal similarity based region merging. Pattern Recogn 43(2):445–456

19. Malmberg F (2011) Graph-based methods for interactive image segmentation. Ph.D. thesis, University West

20. Shi R, Liu Z, Xue Y, Zhang X (2011) Interactive object segmentation using iterative adjustable graph cut. In: Visual communications and image processing (VCIP), IEEE, 2011, pp 1–4

21. Calderero F, Marques F (2010) Region merging techniques using information theory statistical measures. IEEE Trans Image Proces 19(6):1567–1586

22. Couprie C, Grady L, Najman L, Talbot H (2009) Power watersheds: a new image segmentation framework extending graph cuts, random walker and optimal spanning forest. In: 2009 IEEE 12th international conference on computer vision, pp 731–738. IEEE

23. Falcão A, Udupa J, Miyazawa F (2000) An ultra-fast user-steered image segmentation paradigm: live wire on the fly. IEEE Trans Med Imag 19(1):55–62

24. Noma A, Graciano A, Consularo L, Bloch I (2012) Interactive image segmentation by matching attributed relational graphs. Pattern Recogn 45(3):1159–1179

25. Collins LM (2006) Byu scientists create tool for "virtual surgery". Deseret Morning News pp 07–31

26. Mortensen EN, Barrett WA (1995) Intelligent scissors for image composition. In: Proceedings of the 22nd annual conference on Computer graphics and interactive techniques, SIGGRAPH '95, pp. 191–198. ACM, New York (1995)

27. Friedland G, Jantz K, Rojas R (2005) Siox: simple interactive object extraction in still images. In: Seventh IEEE international symposium on multimedia, p 7. IEEE

28. Friedland G, Lenz T, Jantz K, Rojas R (2006) Extending the siox algorithm: alternative clustering methods, sub-pixel accurate object extraction from still images, and generic video segmentation. Free University of Berlin, Department of Computer Science, Technical report B-06-06

29. Gimp G (2008) Image manipulation program. User manual, Edge-detect filters, Sobel, The GIMP Documentation Team

30. Lombaert H, Sun Y, Grady L, Xu C (2005) A multilevel banded graph cuts method for fast image segmentation. In: Tenth IEEE international conference on computer vision, 2005. ICCV, vol 1, pp 259–265. IEEE

31. McGuinness K, OConnor NE (2011) Toward automated evaluation of interactive segmentation. Comput Vis Image Underst 115(6):868–884

32. Boykov Y, Kolmogorov V (2004) An experimental comparison of min-cut/max-flow algorithms for energy minimization in vision. IEEE Trans Pattern Anal Mach Intel 26(9):1124–1137

33. Gauch J, Hsia C (1992) Comparison of three-color image segmentation algorithms in four color spaces. In: Applications in optical science and engineering, pp 1168–1181. International Society for Optics and Photonics
34. Martin D, Fowlkes C, Tal D, Malik J (2001) A database of human segmented natural images and its application to evaluating segmentation algorithms and measuring ecological statistics. In: Proceeding of 8th international conference computer vision, vol 2, pp. 416–423

Chapter 2
Interactive Segmentation: Overview and Classification

Keywords Graph modeling · Markov random field · Maximum a posteriori · Boundary tracking · Label propagation

Being different from automatic image segmentation, interactive segmentation allows user interaction in the segmentation process by providing an initialization and/or feedback control. A user-friendly segmentation system is required in practical applications. Many recent developments have driven interactive segmentation techniques to be more and more efficient. We give an overview on the design of interactive segmentation systems, commonly-used graphic models and classification of segmentation techniques in this chapter.

2.1 System Design

A functional view of an interactive image segmentation system is depicted in Fig. 2.1. It consists of the following three modules:

- User Input Module (Step 1)
  This module receives user input and/or control signals, which helps the system recognize user intention.
- Computational Module (Step 2)
  This is the main part of the system. The segmentation algorithm runs automatically according to user input and generates intermediate segmentation results.
- Output Display Module (Step 3)
  The module delineates and displays the intermediate segmentation results.

The above three steps operate in a loop [1]. In other words, the system allows additional user feedback after Step 3, and then it is back to Step 1. The system runs iteratively until the user gets a satisfied result and terminates the process.

J. He et al., *Interactive Segmentation Techniques*,
SpringerBriefs in Signal Processing
DOI: 10.1007/978-981-4451-60-4_2, © The Author(s) 2014

Fig. 2.1 Illustration of
an interactive image
segmentation system [1],
where a user can control the
process iteratively until a
satisfactory result is obtained

Fig. 2.2 The process of
an interactive segmentation
system

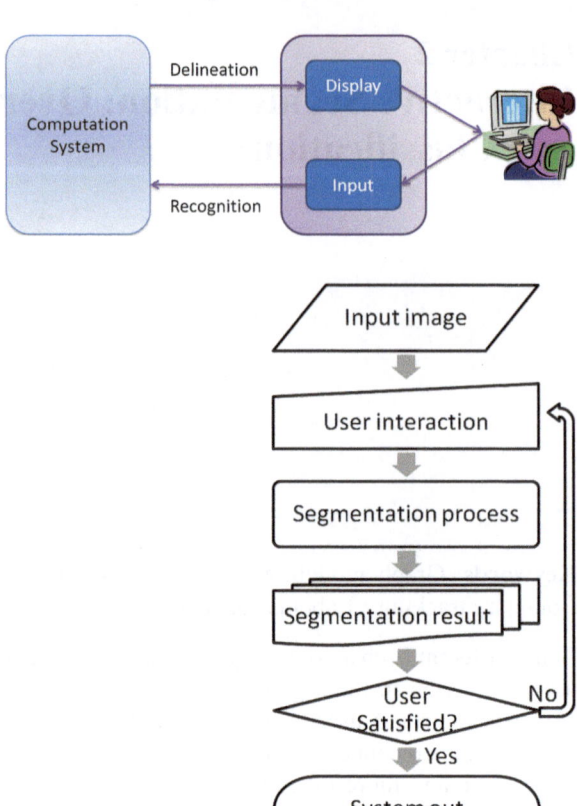

The process of an interactive image segmentation system is shown in Fig. 2.2, where the segmentation objectives rely on human intelligence. Such knowledge is offered to the system via human interaction, which is represented in the form of drawing that provides the color, texture, location, and size information. The interactive segmentation system attempts to understand user intention based on the high-level information so that it can extract the accurate object regions and boundaries eventually. In the process, the system may update and improve the segmentation results through a series of interactions. Thus, it is a human–machine collaborative methodology. On one hand, a machine has to interpret user input and segments the image through an algorithmic process. On the other hand, a user should know how his/her input will affect machine behavior to save the iterations.

There are several user interaction types, and all of them aim to offer the information about background or foreground regions (e.g., brightness, color, location, and size). A user can make strokes to label the object and the background in an image, mark rectangles to locate the object target range, or make control points or seed points to track object boundaries. User interactions vary according to the segmentation algorithms, and the control can be soft or hard constraints in the algorithms [2].

According to Grady [3], an ideal interactive segmentation system should satisfy the following four properties:

- Fast computation in the computational module;
- Fast and easy editing in the user input module;
- Ability to generate an arbitrary segmentation contour given sufficient user control;
- Ability to provide intuitive segmentation results.

Research on interactive segmentation system design has focused on enhancing these four properties. Fast computation and user-friendly interface modules are essential requirements for a practical interactive segmentation system, since an user should be able to sequentially add or remove strokes/marks based on updated segmentation results in real time. This implies that the computational complexity of an interactive image segmentation algorithm should be at an acceptable level. Furthermore, the system should be capable of generating desired object boundaries accurately with a minimum amount of user effort.

2.2 Graph Modeling and Optimal Label Estimation

Image segmentation aims to segment out user's desired objects. Target may consist of multiple homogeneous regions whose pixels share some common properties while the discontinuity of brightness, colors, contrast and texture of image pixels indicates the location of object boundaries [4]. Segmentation algorithms have been developed using the features of pixels, the relationship between pixels and their neighbors, etc. To study these features and connections, one typical approach is to model an input image as a graph, where each pixel corresponds to a node [5].

A graph, denoted by $G = (V, E)$, is a data structure consisting of a set of nodes V and a set of edges E connecting those nodes [5]. If an edge, which is also referred to as a link, has a direction, the graph is called a directed graph. Otherwise, it is an undirected graph. We often model an image as an undirected graph and use a node to represent a pixel in the image. Since each pixel in an image is connected with its neighbors, such as 4-connected neighborhood or 8-connected neighborhood as shown in Fig. 2.3, its associated graph model is structured in the same way. An edge between two nodes represents the connection of these two nodes. In some cases, we may treat an over-segmented region as a basic unit, called the superpixel [6–9], and use a node to represent the superpixel.

An implicit assumption behind the graph approach is that, for a given image I, there exists a probability distribution that can capture labels of nodes and their relationship [10]. Specifically, let node x in graph G be associated with random variable l from set L, which indicates its segmentation status (foreground object or background), the problem of segmenting image I is equivalent to a labeling problem of graph G.

It is often assumed that the graph model of an image satisfies the following Markov properties of conditional independence [5, 10].

(a)   **(b)**

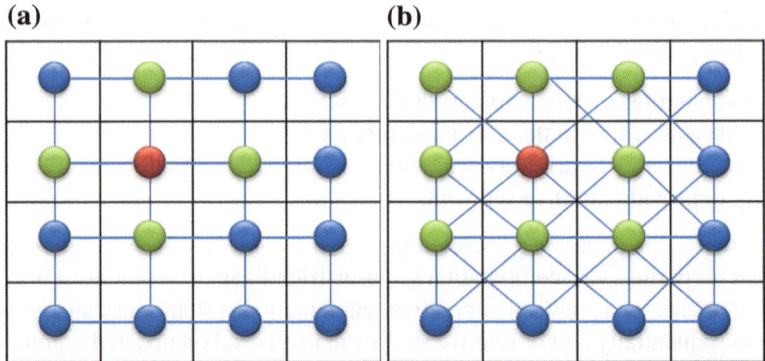

Fig. 2.3 A simple graph example of a 4 × 4 image. The *red* node has 4 *green* neighbors in (**a**) and 8 *green* neighbors in (**b**). **a** Graph with 4-connected neighborhood. **b** Graph with 8-connected neighborhood

- Pairwise Markov independence.
- Any two non-adjacent variables are conditionally independent given all other variables. Mathematically, for non-adjacent nodes x_i and x_j (i.e. $e_{i,j} \notin E$), label l_i and l_j are independent when conditioned on all other variables:

$$Pr(l_i, l_j | L_{V \setminus \{i,j\}}) = Pr(l_i | L_{V \setminus \{i,j\}}) Pr(l_j | L_{V \setminus \{i,j\}}). \tag{2.1}$$

- Local Markov independence.
  Given the neighborhood $\mathcal{N}$ of node x_i, its label l_i is independent of the rest of other labels:

$$Pr(l_i | L_{V \setminus \{i\}}) = Pr(l_i | L_{\mathcal{N}(i)}). \tag{2.2}$$

- Global Markov property.
- Subsets A and B of L are conditionally independent given a separating subset S, where every path connecting a node in A and a node in B passes through S.

If the above properties hold, the graph of image I can be modeled as a Markov random field (MRF) under the Bayesian framework. Figure 2.4 shows the MRF of a 4 × 4 image with 4-connected neighborhood.

Since image segmentation can be formulated as a labeling problem in an MRF, the task becomes the determination of optimal labels for nodes in the MRF. Some node labels are set through user interactions in interactive image segmentation. With the input image as well as this prior knowledge, the maximum a posteriori (MAP) method provides an effective solution to the label estimation of the remaining nodes in the graph [5, 10, 11]. According to the Bayesian rule, the posterior probability of node labels can be written as

$$Pr(l_{1 \ldots N} | x_{1 \ldots N}) = \frac{\prod_{i=1}^{N} Pr(x_i | l_i) Pr(l_{1 \ldots N})}{Pr(x_{1 \ldots N})}, \tag{2.3}$$

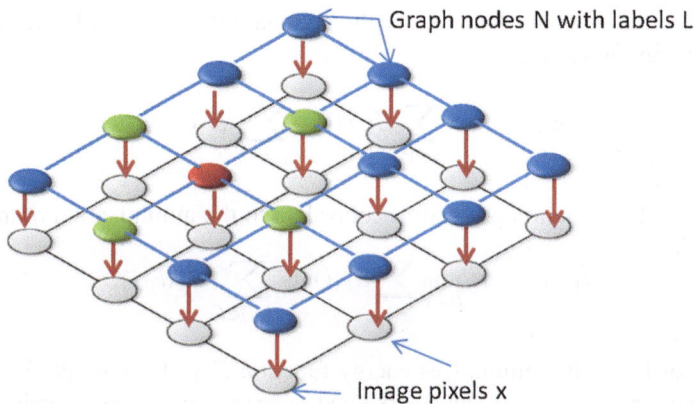

Graph nodes N with labels L

Image pixels x

Fig. 2.4 The MRF model of a 4 × 4 image

where $Pr(l)$ is the prior probability of labels and $Pr(x|l)$ is the conditional probability of the node value conditioned on a certain label.

In the MAP estimation, we search for labels $\hat{l}_{1...N}$ that maximize the posterior probability:

$$\hat{l}_{1...N} = \arg\max_{l_{1...N}} Pr(l_{1...N}|x_{1...N})$$

$$= \arg\max_{l_{1...N}} \prod_{i=1}^{N} Pr(x_i|l_i)Pr(l_{1...N})$$

$$= \arg\max_{l_{1...N}} \sum_{i=1}^{N} \log[Pr(x_i|l_i)] + \log[Pr(l_{1...N})] \tag{2.4}$$

By using the pairwise Markov property and the clique potential of the MRF, we have

$$Pr(l_{1...N}) \propto \exp(-\sum_{(i,j)\in E} E_2(i,j)), \tag{2.5}$$

where E_2 is the pairwise clique potential. It can be regarded as an energy function to measure the compatibility of neighboring labels [5]. In the segmentation literature, it is also called the smoothness term or the boundary term because of its physical meaning in the segmentation [1, 12, 13].

The first term in Eq. (2.4) is a log likelihood function. By following Eq. (2.4) and with the following association

$$E_1(i) = -\log[Pr(x_i|l_i)], \tag{2.6}$$

which is referred to as the data term or the regional term [1, 12, 13], we can define
the total energy function as

$$E(l_{1\ldots N}) = \sum_{i \in V} E_1(i) + \sum_{(i,j) \in E} E_2(i, j). \qquad (2.7)$$

Then, the MAP estimation problem is equivalent to the minimization of the energy
function E:

$$\hat{l}_{1\ldots N} = \arg\min_{l_{1\ldots N}} \sum_{i \in V} E_1(i) + \sum_{(i,j) \in E} E_2(i, j). \qquad (2.8)$$

The set L of labels that minimizes energy function E yields the optimal solution.
Finally, segmentation results can be obtained by extracting pixels or regions associated with the foreground labels. We will present several methods to solve the above
energy minimization problem in Chap. 3.

2.3 Classification of Solution Techniques

One can classify interactive segmentation techniques into several types based on
different viewpoints. They are discussed in detail below.

- **Application-Driven Segmentation**
  One way to classify interactive segmentation methods is based on their applications. Some techniques target at generic natural image segmentation, while others
  address medical and industrial image applications as shown in Fig. 2.5. Natural
  images tend to have rich color information, which offer a wide range of intensity features for the segmentation purpose. The challenges lie in weak boundaries
  and ambiguous objects. An ideal segmentation method should be robust in handling a wide variety of images consisting of various color intensity, luminance,
  object size, object location, etc. Furthermore, no shape prior is used. In contrast,
  for image segmentation in medical and industrial applications, most images are
  monochrome and segmentation objects are often specific such as cardiac cells [14],
  neuron [15], and brain and born CT [3]. Without the color information, medical
  image segmentation primarily relies on the luminance information. It is popular

Fig. 2.5 Application-
driven interactive image
segmentation algorithms

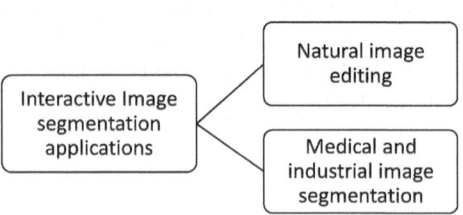

to learn specific object shape priors to add more constraints so as to cut out the desired object.

- **Mathematical Models and Tools**

  Another way to classify image segmentation methods is based on the mathematical models and tools. Graph models were presented in Sect. 2.2. Other models and tools include Bayesian theory, Gaussian mixture models [6], Gaussian mixture Markov random fields [16], Markov random fields [10], random walks [3], random walks with restart [17], min-cut/max-flow [12, 18], cellular automation [19], belief propagation [20], and spline representation [21].

- **Form of Energy Function**

  Another viewpoint is to consider the form of the energy function. Generally, the energy function consists a data term and a smoothness term. The data term represents labeled image pixels or superpixels of objects and background defined through user interactions. The smoothness term specifies local relations between pixels or super pixels with their neighborhood (such as color similarity and the edge information). The combined energy function should strike a balance between those two terms. The mathematical tool for energy minimization and the similarity measure also provides a categorization tool for segmentation methods.

- **Local Versus Global Optimization**

  It is typical to adopt one of the following two approaches to minimize the energy function. One is to find the local discontinuity of the image amplitude attribution to locate the boundary of a desired object and cut it out directly. The other is to model the image as a probabilistic graph and use some prior information (such as user scribbles, shape priors, or even learned models) to determine the labels of remaining pixels under some global constraints. Commonly used operations include classification, clustering, splitting and merging, and optimization. The global constraints are used to generate smooth and desired object regions by connecting neighboring pixels of similar attributes and separating pixels in discontinuous regions (i.e., boundary regions between objects and the background) as much as possible. Here, a graph model is built to capture not only the color features of pixels but also the spatial relationship between adjacent pixels. The global optimization should consider these two kinds of relations. Image segmentation approaches also vary on the definitions of the attributions of the probabilistic graphs and the measurements of feature similarities between local connected nodes.

- **Pixel-wise Versus Region-wise Processing**

  Pre-processing and post-processing techniques can be used to speedup the computation time and improve segmentation accuracy [6–8]. Pixelwise processing has the potential to yield more accurate results along object boundaries. However, it may not be necessary to apply it to all pixels of an image since pixels in homogeneous regions have high similarity and their segmentation labels are likely the same. Based on this observation, it is possible to apply a pre-segmentation method that merges locally homogeneous pixels into local regions. This results in over-segmentation where each local region can be treated as a superpixel [6–8] and modeled as a node in a graph model as shown in Fig. 2.6. This strategy reduces

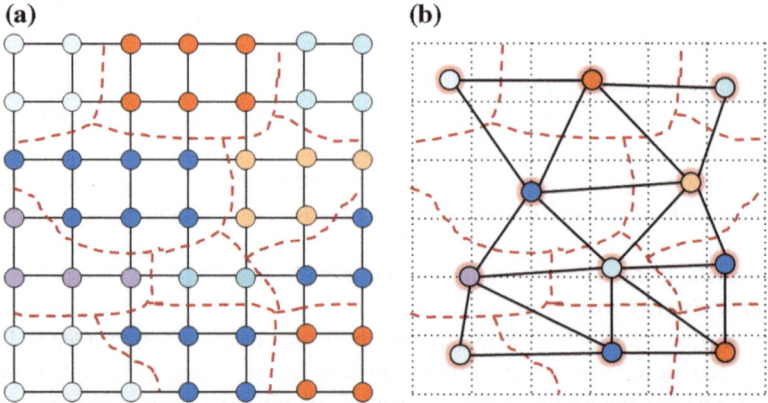

Fig. 2.6 A graph model for a 7×7 image with over-segmentation. Homogeneous pixels are merged into a superpixel, and then represented as a node in the graph in (**b**), where edges connect neighboring superpixels in (**a**). **a** Image with oversegmentation. **b** Graph based on superpixels

the processing time of large images significantly. On the other hand, it may cause errors near object boundaries. Since the segmentation is based on superpixels, it is challenging to locate accurate boundaries. For objects with soft boundaries, a hard boundary segmentation is not acceptable. Then, a post-segmentation procedure is needed to refine object boundaries. Image matting [6, 8, 20, 22] can be viewed as a post-processing technique as well.

- **Boundary Tracking Versus Label Propagation**
  Based on interaction types, interactive image segmentation algorithms can be classified into two main categories as shown in Fig. 2.7: (1) tracking the object boundary, and (2) propagating segment labels . To track the object boundary, a user can move the cursor along the object boundary so that the system can find boundary contours based on cursor movement. Once the boundary is closed as a loop, an object can be extracted along the boundary. For example, a user can move the cursor to conduct online segmentation in intelligent scissors [23, 24]. In this category, one focuses on the location of boundaries rather than on the optimization of an energy function. An alternative approach is to propagate user labels from parts of the object and background of an input image to the remaining parts. Then, the image can be segmented into two or multiple partitions as object and background according to the label of each pixel. There are several ways to propagate user labels, such as graph-cut [25], region-based splitting and merging [26], and graph matching [8]. Segmentation accuracy varies depending on the efficiency of label propagation. Due to different algorithmic complexities, user interactions can be done offline or online. Examples include offline stroking and marking a rectangular object region offline [13, 25] or online interactions [6, 8]. The trend is to move from offline to online interactions by either lowering the complexity or enhancing the computational power of the machine.

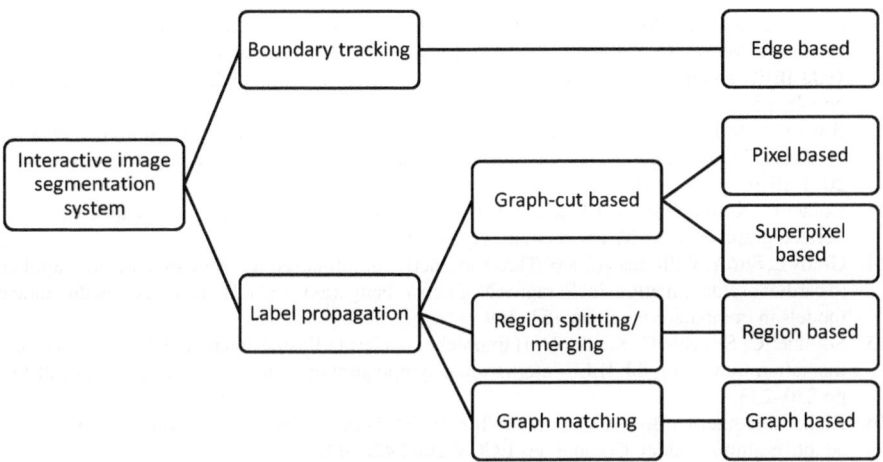

Fig. 2.7 Classification of interactive segmentation methods based on interaction types

As an extension, an interactive segmentation system can segment multiple objects at once [8]. It is also possible for an algorithm to conduct the segmentation task on multiple similar images at once [8, 27] and even for video segmentation [22]. Furthermore, one often encounters 3D (or volumetric) images in the context of medical imaging. Some techniques have been generalized from the 2D to the 3D case for medical image segmentation [21, 25]. In this book, we focus on the 2D image segmentation.

References

1. Malmberg F (2011) Graph-based methods for interactive image segmentation. Ph.D. thesis, University West
2. Yang W, Cai J, Zheng J, Luo J (2010) User-friendly interactive image segmentation through unified combinatorial user inputs. IEEE Trans Image Process 19(9):2470–2479
3. Grady L (2006) Random walks for image segmentation. IEEE Trans Pattern Anal Mach Intell 28(11):1768–1783
4. Pratt W (2007) Digital image processing: PIKS scientific inside. Wiley-Interscience publication. Hoboken, NJ, USA
5. Koller D, Friedman N (2009) Probabilistic graphical models: principles and techniques. MIT press, Cambridge, MA, USA
6. Li Y, Sun J, Tang CK, Shum HY (2004) Lazy snapping. ACM Trans Graph 23(3):303–308
7. Ning J, Zhang L, Zhang D, Wu C (2010) Interactive image segmentation by maximal similarity based region merging. Pattern Recogn 43(2):445–456
8. Noma A, Graciano A, Consularo L, Bloch I (2012) Interactive image segmentation by matching attributed relational graphs. Pattern Recogn 45(3):1159–1179
9. Ren X, Malik J (2003) Learning a classification model for segmentation. In: Proceedings of the 9th IEEE international conference on computer vision, vol 2, ICCV '03. IEEE Computer Society, Washington, DC, USA, pp 10–17

10. Perez P et al (1998) Markov random fields and images. CWI quarterly 11(4):413–437
11. Boykov Y, Veksler O, Zabih R (1998) Markov random fields with efficient approximations. In: 1998 IEEE computer society conference on computer vision and pattern recognition. IEEE, pp 648–655
12. Boykov Y, Jolly M (2001) Interactive graph cuts for optimal boundary and region segmentation of objects in nd images. In: Eighth IEEE international conference on computer vision, vol 1 2001. IEEE, pp 105–112
13. Rother C, Kolmogorov V, Blake A (2004) "grabcut": interactive foreground extraction using iterated graph cuts. ACM Trans Graph 23(3):309–314
14. Grady L, Sun Y, Williams J (2006) Three interactive graph-based segmentation methods applied to cardiovascular imaging. In: Paragios N, Chen Y, Faugeras O (eds) Handbook of mathematical models in computer vision, pp 453–469
15. Sommer C, Straehle C, Koethe U, Hamprecht FA (2011) Ilastik: interactive learning and segmentation toolkit. In: 8th IEEE international symposium on biomedical imaging (ISBI 2011): pp 230–233
16. Blake A, Rother C, Brown M, Perez P, Torr P (2004) Interactive image segmentation using an adaptive gmmrf model. Comput Vis ECCV 2004:428–441
17. Kim T, Lee K, Lee S (2008) Generative image segmentation using random walks with restart. Comput Vis ECCV 2008:264–275
18. Boykov Y, Veksler O (2006) Graph cuts in vision and graphics: theories and applications. In: Paragios N, Chen Y, Fangeras O (eds) Handbook of mathematical models in computer vision, pp 79–96
19. Vezhnevets V, Konouchine V (2005) Growcut: interactive multi-label nd image segmentation by cellular automata. In: Proceedings of graphicon, pp 150–156
20. Wang J, Cohen MF (2005) An iterative optimization approach for unified image segmentation and matting. In: Tenth IEEE international conference on computer vision, vol 2, ICCV 2005. IEEE, pp 936–943
21. Kass M, Witkin A, Terzopoulos D (1988) Snakes: Active contour models. Int J Comput Vis 1(4):321–331
22. Bai X, Sapiro G (2007) A geodesic framework for fast interactive image and video segmentation and matting. In: IEEE 11th international conference on computer vision, 2007. IEEE, pp 1–8
23. Barrett W, Mortensen E (1997) Interactive live-wire boundary extraction. Med Image Anal 1(4):331–341
24. Mortensen EN, Barrett WA (1995) Intelligent scissors for image composition. In: Proceedings of the 22nd annual conference on computer graphics and interactive techniques, SIGGRAPH '95. ACM, New York, NY, USA, pp 191–198
25. Boykov Y, Funka-Lea G (2006) Graph cuts and efficient n-d image segmentation. Int J Comput Vis 70(2):109–131
26. Adams R, Bischof L (1994) Seeded region growing. IEEE Trans Pattern Anal Mach Intell 16(6):641–647
27. Batra D, Kowdle A, Parikh D, Luo J, Chen T (2010) Icoseg: interactive co-segmentation with intelligent scribble guidance. In: IEEE conference on computer vision and pattern recognition (CVPR) 2010. IEEE, pp 3169–3176

Chapter 3
Interactive Image Segmentation Techniques

Keywords Graph-cut · Random walks · Active contour · Matching attributed relational graph · Region merging · Matting

Interactive image segmentation techniques are semiautomatic image processing approaches. They are used to track object boundaries and/or propagate labels to other regions by following user guidance so that heterogeneous regions in one image can be separated. User interactions provide the high-level information indicating the "object" and "background" regions. Then, various features such as locations, color intensities, local gradients can be extracted and used to provide the information to separate desired objects from the background. We introduce several interactive image segmentation methods according to different models and used image features.

This chapter is organized as follows. First, we introduce several popular methods based on the common graph-cut model in Sect. 3.1. Next, we discuss edge-based, live-wire, and active contour methods that track object boundaries in Sect. 3.2, and examine methods that propagate pixel/region labels by random walks in Sect. 3.3. Then, image segmentation methods based on clustered regions are investigated in Sect. 3.4. Finally, a brief overview of the boundary refinement technique known as matting is offered in Sect. 3.5.

3.1 Graph-Cut Methods

Boykov and Jolly [1] first proposed a graph-cut approach for interactive image segmentation in 2001. They formulated the interactive segmentation as a maximum a posteriori estimation problem under the Markov random field (MAP-MRF) framework [2], and solved the problem for a globally optimal solution by using a fast min-cut/max-flow algorithm [3]. Afterwards, several variants and extensions such

J. He et al., *Interactive Segmentation Techniques*,
SpringerBriefs in Signal Processing
DOI: 10.1007/978-981-4451-60-4_3, © The Author(s) 2014

as GrabCut [4] and Lazy Snapping [5] have been developed to make the graph-cut approach more efficient and easier to use.

3.1.1 Basic Idea

In interactive segmentation, we expect a user to provide hints about objects that are to be segmented out from an input image. In other words, a user provides the information to meet the segmentation objectives. For example, in Boykov and Jolly's work [1], a user marks certain pixels as either the "object" or the "background," which are referred to as seeds, to provide hard constraints for the later segmentation task. Then, a graph-cut optimization procedure is performed to obtain a globally optimum solution among all possible segmentations that satisfy these hard constraints. At the same time, boundary and region properties are incorporated in the cost function of the optimization problem, and these properties are viewed as soft constraints for segmentation.

As introduced in Sect. 2.2, Boykov et al. [1, 7] defined a directional graph $G = \{V, E\}$, which consists of a set, V, of nodes (or vertices) and a set, E, of directed edges that connect nodes. In interactive segmentation, user seeded pixels for objects and background are, respectively, represented by source node s and sink node t. Each unmarked pixel is associated with a node in the 2D plane. As a result, V consists of two terminal nodes, s and t, and a set of non-terminal nodes in graph G which is denoted by I. We connect selected pairs of nodes with edges and assign each edge a non-negative cost. The edge cost from node x_i and node x_j is denoted as $c(x_i, x_j)$. In a directed graph, the edge cost from x_i to x_j is in general different from that from x_j to x_i. That is,

$$c(x_i, x_j) \neq c(x_j, x_i). \tag{3.1}$$

Figure 3.1b shows a simple graph with terminal nodes s and t and non-terminal nodes x_i and x_j.

An edge is called a t-link, if it connects a non-terminal node in I to terminal node t or s. An edge is called a n-link, if it connects two non-terminal nodes in I. Let F be the set of n-links. One can partition E into two subsets F and $E - F$ [7], where

$$E - F = \{(s, x_i), (x_j, t), \forall x_i, x_j \in I\}. \tag{3.2}$$

In Fig. 3.1, t-links are shown in black while n-links are shown in red.

A cut $C \subset E$ partitions vertices in a graph into two disjoint subsets S and T, where source node s belongs to S and sink node t belongs to T. Figure 3.1b shows an example of a cut. Typically, a cost function is used to measure the efficiency of a cut. The weight of an n-link represents a penalty for discontinuity between its connecting nodes while the weight of a t-link indicates the labeling cost to associate a non-terminal node to the source or the sink [8]. The cost of a cut is the sum of weights of edges severed by the cut. The optimal cut C minimizes the cut cost.

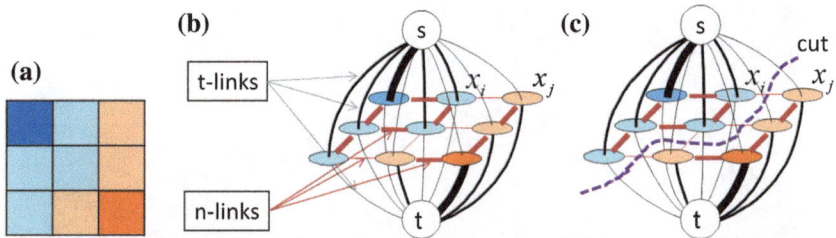

Fig. 3.1 A simple graph example of a 3 × 3 image, where all nodes are connected to source node s and sink node t and edge thickness represents the strength of the connection [6]. **a** 3 × 3 image. **b** A graph model for (**a**). **c** A cut for (**a**)

Ford and Fulkerson's theorem [9] states that minimizing the cut cost is equivalent to maximizing the flow through the graph network from source node s to sink node t. The corresponding cut is called a minimum cut. Thus, finding a minimum cut is equivalent to solving the max-flow problem in the graph network. Many algorithms have been developed to solve the min-cut/max-flow problem, e.g., [2, 10]. One may use any of the algorithms to obtain disjoint sets S and T and label all pixels, corresponding to nodes in S, as the "foreground object" and label the remaining pixels, corresponding to nodes in T, as the "background". Then, the segmentation task is completed.

3.1.2 Interactive Graph-Cut

Boykov and Jolly [1] proposed an interactive graph-cut method, where a user indicates the locations of source and sink pixels. The problem was cast in the MAP-MRF framework [2]. A globally optimal solution was derived and solved by a fast min-cut/max-flow algorithm. The energy function is defined as

$$E(L) = \sum_{i \in V} D_i(L_i) + \sum_{(i,j) \in E} V_{i,j}(L_i, L_j), \qquad (3.3)$$

where $L = \{L_i | x_i \in I\}$ is a binary labeling scheme for image pixels (i.e., $L_i = 0$ if x_i is a background pixel and $L_i = 1$ if x_i is a foreground object pixel), $D_i(\cdot)$ is a pre-specified likelihood function used to indicate the labeling preference for pixels based on their colors or intensities, $V_{i,j}(\cdot)$ denotes a boundary cost, and $(i, j) \in E$ means that x_i and x_j are adjacent nodes connected by edge (x_i, x_j) in graph G. The boundary cost, $V_{i,j}$, encourages the spatial coherence by penalizing the cases where adjacent pixels have different labels [8]. Normally, the penalty gets larger, when x_i and x_j are similar in colors or intensities, and it approaches zero when the two pixels are very different. The similarity between x_i and x_j can be measured in many ways (e.g., local intensity gradients, Laplacian zero-crossing or gradient directions). Note

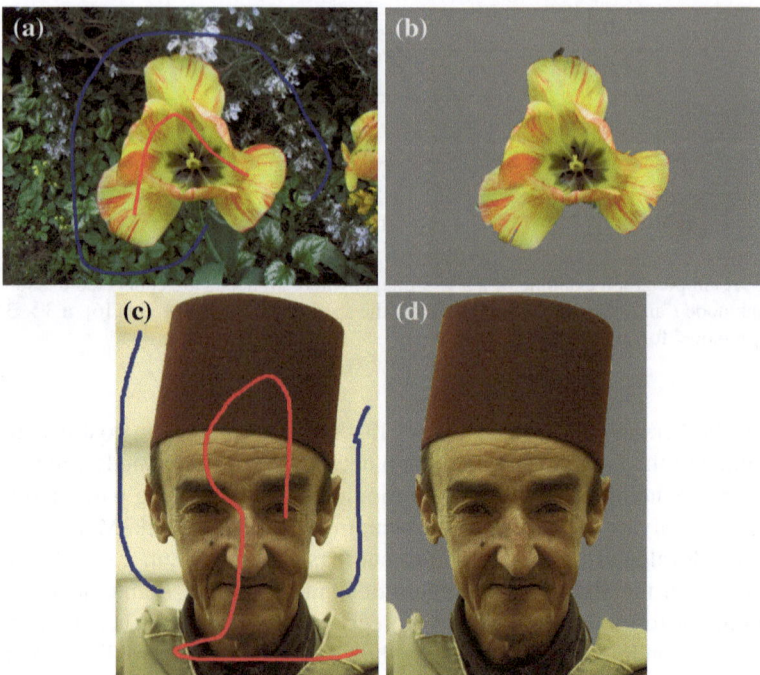

Fig. 3.2 Two segmentation results obtained by using the interactive graph-cut algorithm [3, 6]. **a** Original image with user markup. **b** Segmentation result of the flower. **c** Original image with user markup. **d** Segmentation result of the man

that this type of energy functions, composed of regional and boundary terms, are employed in most graph-based segmentation algorithms.

The interactive graph-cut algorithm often uses a multiplier $\lambda \geq 0$ to specify a relative importance of regional term D_i in comparison with boundary term $V_{i,j}$. Thus, we rewrite Eq. (3.3) as:

$$E(L) = \lambda \cdot \sum_{x_i \in I} D_i(L_i) + \sum_{(x_i, x_j) \in N} V_{i,j}(L_i, L_j). \qquad (3.4)$$

The intensities of marked seed pixels are used to estimate the intensity distributions of foreground objects and background regions, denoted as $Pr(I|F)$ and $Pr(I|B)$, respectively. Being motivated by [2], the interactive graph-cut algorithm defines the regional term with negative log-likelihoods in the following form:

$$D_I(L_i = 1) = -\ln Pr(I(i)|F), \qquad (3.5)$$

$$D_I(L_i = 0) = -\ln Pr(I(i)|B), \qquad (3.6)$$

where $I(i)$ is the intensity of pixel x_i and $Pr(I(i)|L)$ can be computed based on the intensity histogram. The boundary term can be defined as:

$$V_{i,j}(L_i, L_j) \propto |L_i - L_j| \exp\left(-\frac{(I(i) - I(j))^2}{2\sigma^2}\right) \cdot \frac{1}{d(i, j)}, \qquad (3.7)$$

where $d(i, j)$ is the spatial distance between pixels x_i and x_j and the deviation, σ, is a parameter related to the camera noise level. The similarity of pixels x_i and x_j is computed based on the Gaussian distribution. Finally, the interactive graph-cut algorithm obtains the labeling (or segmentation) result L by minimizing the energy function in (3.4).

Figure 3.2 shows two segmentation results of the interactive graph-cut algorithm, where the red strokes indicate foreground objects while the blue strokes mark the background region to model the intensity distributions. Segmentation results are obtained by minimizing the cost function in (3.4).

3.1.3 GrabCut

Rother et al. [4] proposed a GrabCut algorihtm by extending the interactive graph-cut algorithm with an iterative process. GrabCut uses the graph-cut optimization procedure as discussed in Sect. 3.1.2 at each iteration. It has three main features.

1. GrabCut uses a Gaussian mixture model (GMM) to represent pixel colors (instead of the monochrome histogram model in the interactive graph-cut algorithm).
2. GrabCut alternates between object estimation and GMM parameter estimation iteratively while the optimization is done only once in the interactive graph-cut algorithm.
3. GrabCut demands less user interaction. Basically, a user only has to place a rectangle or lasso around an object (instead of detailed strokes) as illustrated in Fig. 3.3. A user can still draw strokes for further refinement if needed.

GrabCut processes a color image in the RGB space. It uses GMMs to model the color distributions of the object and background, respectively. Each GMM is trained to be a full-covariance Gaussian mixture with K components. Let $\mathbf{k} = (k_1, \ldots, k_n, \ldots, k_N)$, $k_n \in \{1, \ldots, K\}$, where subscript n denotes the pixel index and N is the total number of pixels within the marked region. Vector $\mathbf{k}$ assigns each pixel a unique GMM component. The object model and the background model of a pixel with index n are denoted by $\alpha_n = 0$ and 1, respectively. Then, the energy function can be written as

$$E(\alpha, \mathbf{k}, \theta, \mathbf{z}) = U(\alpha, \mathbf{k}, \theta, \mathbf{z}) + V(\alpha, \mathbf{z}), \qquad (3.8)$$

where $\mathbf{z}$ is the image data, θ represents the GMM model parameters,

Fig. 3.3 Segmentation results of GrabCut, which requires a user to simply place a rectangle around the object of interest. **a** Original image with *rectangle* markup. **b** Segmentation result of the flower and butterfly. **c** Original image with *rectangle* markup. **d** Segmentation result of the kid

$$\theta = \{\pi(\alpha, k), \mu(\alpha, k), \Sigma(\alpha, k)\} \tag{3.9}$$

where $\alpha = 0, 1, k = 1, \ldots, K$; π is the mixing weight, and μ and Σ are the mean and the covariance matrix of a Gaussian component. The data term is given by

$$U(\alpha, \mathbf{k}, \theta, \mathbf{z}) = \sum_n D(\alpha_n, k_n, \theta_n, z_n), \tag{3.10}$$

where

$$D(\alpha_n, k_n, \theta, z_n) = -\log \pi(\alpha_n, k_n) + \frac{1}{2} \log |\Sigma(\alpha_n, k_n)|$$
$$+ \frac{1}{2}(z_n - \mu(\alpha_n, k_n)])^T (\Sigma(\alpha_n, k_n))^{-1}(z_n - \mu(\alpha_n, k_n)) \tag{3.11}$$

The smoothness term V in (3.8) is computed using the Euclidean distance in the RGB color space,

$$V(\alpha, \mathbf{z}) = \gamma \cdot \sum_{(m,n) \in E} \Psi(\alpha_n \neq \alpha_m) \exp\left(-\beta \|z_m - z_n\|^2\right), \qquad (3.12)$$

where $\Psi(\cdot)$ is the indicator function that has value 1 if the statement is true and 0, otherwise.

The system first assumes an initial segmentation result by choosing membership vectors $\mathbf{k}$ and α. Then, it determines the GMM parameter vector θ by minimizing the energy function in (3.8). Afterwards, with fixed parameter vector θ, it refines the segmentation result α and the Gaussian component membership $\mathbf{k}$ by also minimizing the energy function in (3.8). The above two steps are iteratively performed until the system converges.

Rother et al. also proposed a border matting scheme in [4] that refines binary segmentation results to become soft results near the boundary strip of fixed width, where the segmented boundaries are smoother.

3.1.4 Lazy Snapping

Li et al. [5] proposed the Lazy Snapping algorithm as an improvement over the interactive graph-cut scheme in two areas—speed and accuracy.

- To enhance the segmentation speed, Lazy Snapping adopts over-segmented super-pixels to construct a graph so as to reduce the number of nodes in the labeling computation. A novel graph-cut formulation is proposed by employing pre-computed image over-segmentation results instead of image pixels. The processing speed is accelerated by about 10 times [5].
- To improve the segmentation accuracy, the watershed algorithm [11], which can locate boundaries in an image well and preserve small differences inside each segment, is used to initialize the over-segmentation in the pre-segmentation stage; it also optimizes the object boundary by maximizing color similarities within the object and gradient magnitudes across the boundary between the object and background.

Figure 3.4 shows pre-segmented superpixels , which are used as nodes for the min-cut formulation.

After watershed's pre-segmentation, an image is decomposed into small regions. Each small region corresponds a node in graph $G = \{V, E\}$. The location and the color of a node are given by the central position and the average color of the corresponding small region, respectively. The cost function is defined as

$$E(X) = \sum_{i \in V} E_1(x_i) + \lambda \cdot \sum_{(i,j) \in E} E_2(x_i, x_j), \qquad (3.13)$$

where label x_i takes a binary value (e.g., 1 or 0 if region i belongs to an object or background, respectively), $E_1(x_i)$ is the likelihood energy used to encode the color

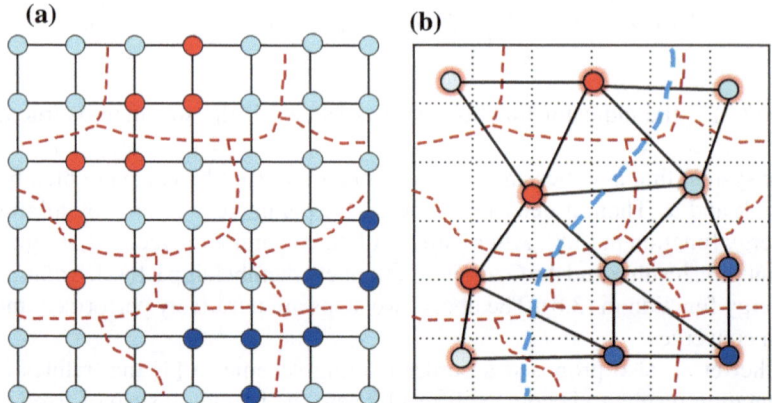

Fig. 3.4 Illustration of the Lazy Snapping algorithm: super-pixels and user strokes (*left*) and each superpixel being converted to a node in the graph (*right*). In this example, *red strokes* stand for the foreground region while *blue strokes* denote the background region. Then, superpixels containing seed pixels are labeled according to their stroke types. The segmentation problem is cast as the labeling of the remaining nodes in the graph. **a** Superpixels and user strokes. **b** Graph with seeded nodes

similarity of nodes and $E_2(x_i, x_j)$ is a penalty term when adjacent nodes are assigned different labels. The terms $E_1(x_i)$ and $E_2(x_i, x_j)$ are defined below.

Each node in the graph represents a small region. Furthermore, we can define foreground seed F and background seed B for some nodes. The colors of F and B are computed by the K-means algorithm, and the mean color clusters are denoted by K_n^F and K_m^B, respectively. Then, the minimum distance from node i with color $C(i)$ to the foreground and background are defined, respectively, as

$$d_i^F = \min_n \| C(i) - K_n^F \|, \text{ and } d_i^B = \min_m \| C(i) - K_m^B \|. \tag{3.14}$$

Then, $E_1(x_i)$ is defined as follows:

$$\begin{cases} E_1(x_i = 1) = 0 & \text{and } E_1(x_i = 0) = \infty, & \text{if } i \in F \\ E_1(x_i = 1) = \infty & \text{and } E_1(x_i = 0) = 0, & \text{if } i \in B \\ E_1(x_i = 1) = \frac{d_i^F}{d_i^F + d_i^B} & \text{and } E_1(x_i = 0) = \frac{d_i^B}{d_i^F + d_i^B}, & \text{otherwise.} \end{cases} \tag{3.15}$$

The prior energy, $E_2(x_i, x_j)$, defines a penalty term when adjacent nodes are assigned with different labels. It is in form of

$$E_2(x_i, x_j) = \frac{|x_i - x_j|}{1 + C_{ij}}, \tag{3.16}$$

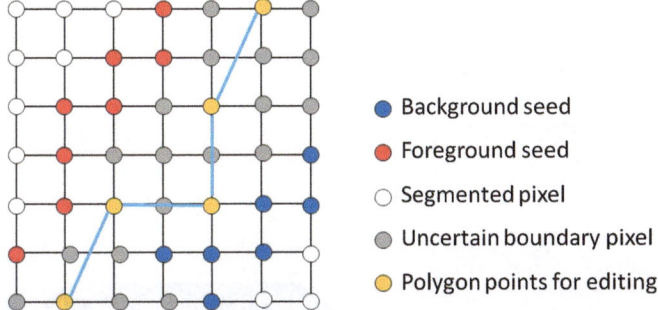

- ● Background seed
- ● Foreground seed
- ○ Segmented pixel
- ● Uncertain boundary pixel
- ● Polygon points for editing

Fig. 3.5 Boundary editing which allows pixel-level refinement on boundaries [5]

where C_{ij} is the mean color difference between regions i and j, which is normalized by the shared boundary length.

Another feature of Lazy Snapping is that it supports boundary editing to achieve pixel-level accuracy as shown in Fig. 3.5. It first converts segmented object boundaries into an editable polygon. Then, it provides two methods for boundary editing:

- Direct vertex editing
  It allows users to adjust the shape of the polygon by dragging vertices directly.
- Overriding brush
  It enables users to add strokes to replace the polygon.

Then, regions around the polygon can be segmented with pixel-level accuracy using the graph-cut. To achieve this objective, the segmentation problem is formulated at the pixel level. The prior energy is redefined using the polygon location as the soft constraint:

$$E_2(x_i, x_j) = \frac{|x_i - x_j|}{1 + (1 - \beta)C_{ij} + \frac{\beta\eta}{D_{ij}^2 + 1}}, \tag{3.17}$$

where x_i is the label for pixel i, D_{ij} is the distance from the center of arc (i, j) to the polygon, η is the scale parameter, and $\beta \in [0, 1]$ is used to balance the influence of D_{ij}. The likelihood energy, E_1, is defined in the same way as (3.15). The final segmentation result is generated by minimizing the energy function. Some segmentation examples can be found in [5] and the website http://youtu.be/WoNwNXkenS4.

3.1.5 Geodesic Graph-Cut

The graph-cut approach sometimes suffers from the problem of short-cutting, which is caused by a lower cost along a shorter cut than that of a real boundary. As shown in

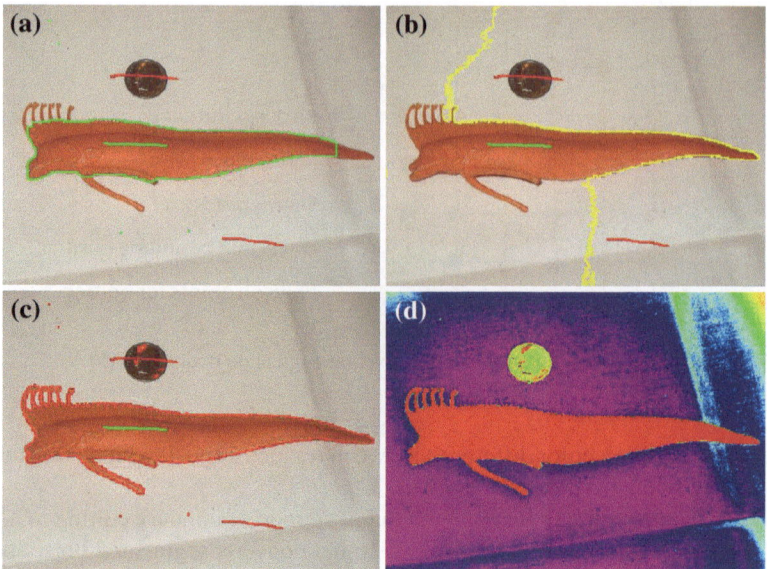

Fig. 3.6 Comparison of segmentation results with the same scribbles as the user input: **a** the short-cutting problem in the standard graph-cut [6]; **b** the false boundary problem in the geodesic segmentation [13]; **c** the geodesic graph-cut [12]; and **d** the geodesic confidence map in [12] to weight between the edge finding and the region modeling

Fig. 3.6, the geodesic graph-cut algorithm [12] attempts to overcome this problem by utilizing the geodesic distance. It also provides users more freedom to place scribbles.

The Euclidean distance between two vertices, $x_i = (x_{i1}, x_{i2})$ and $x_j = (x_{j1}, x_{j2})$, is defined as the l-2 norm of vector $v_{i,j}$ that connects x_i and x_j:

$$d_{i,j} = ||v_{i,j}||_2 = \sqrt{(x_{i1} - x_{j1})^2 + (x_{i2} - x_{j2})^2}. \qquad (3.18)$$

The Euclidean distance, which is often used in the graph-cut algorithm, computes the color similarity, e.g., in Eq. (3.7), without taking other properties of pixels along the path into consideration. The geodesic distance between vertices x_i and x_j is defined as the lowest cost of the transfering path between them, where the cost between two adjacent pixels may vary depending on several factors. If there is no path connecting vertices x_i and x_j, the geodesic distance between them is infinite. The data term in the standard graph-cut algorithm is typically calculated based on the log-likelihood of the color histogram without considering factors such as the locations of object boundaries and seeded points. In contrast, the geodesic graph-cut method uses the geodesic distance as one of the data terms.

Each seed pixel s is either labeled as foreground (F) or background (B). We use Ω_l to denote the set of labeled seed pixels with label $l \in \{F, B\}$ and $d_l(x_i, x_j)$ to denote the geodesic distance from pixel x_i to pixel x_j based on a color model and Ω_l. Then, $d_l(x_i, x_j)$ is defined to be the minimum cost among all paths, C_{x_i,x_j}, connecting x_i and x_j. Mathematically, we have

$$d_l(x_i, x_j) = \min_{C_{x_i,x_j}} \int_0^1 |W_l \cdot \dot{C}_{x_i,x_j}(p)| dp, \qquad (3.19)$$

where W_l are weights along path C_{x_i,x_j}. Often, W_l is set to the gradient of the likelihood that pixel x on this path belongs to the foreground; namely

$$W_l(x) = \nabla P_l(x), \qquad (3.20)$$

where

$$P_l(x) = \frac{P_r(c(x)|l)}{P_r(c(x)|F) + P_r(c(x)|B)}, \qquad (3.21)$$

and where $c(x)$ is the color of pixel x and $P_r(c(x)|l)$ is the probability of color $c(x)$ given a color model and Ω_l. Then, the geodesic distance $D_l(x_i)$ of pixel x_i is defined as the smallest geodesic distance $d_l(s, x_i)$ from pixel x_i to each seed pixel in form of

$$D_l(x_i) = \min_{s \in \Omega_l} d_l(s, x_i). \qquad (3.22)$$

Finally, pixel x_i will be labeled with the same label of the nearest seed pixels measured by the geodesic distance.

Bai et al. [13] extended the above solution to the soft segmentation problem, in which the alpha matte, $\alpha(x)$, for each pixel x is computed explicitly via

$$\alpha(x) = \frac{w_F(x)}{w_F(x) + w_B(x)}, \qquad (3.23)$$

where

$$w_l(x) = D_l(x)^{-r} \cdot P_l(x), \quad l \in \{F, B\}. \qquad (3.24)$$

For the hard segmentation problem, the foreground object has $\alpha = 1$ while the background region has $\alpha = 0$. The final segmentation results can be obtained by extracting regions with $\alpha = 1$. Sometimes, a threshold for α is set to extract parts of the translucent boundaries along with solid foreground objects.

There are several other geodesic graph-cut algorithms. For example, based on a similar geodesic distance defined in [13], Criminisi et al. [14] computed the geodesic distance to offer a set of sensible and restricted possible segments, and obtained an optimal segmentation by finding the solution that minimizes the cost energy. Being different from the conventional global energy minimization, Criminisi et al. [14] addressed this problem by finding a local minimum.

Another example is the geodesic graph-cut algorithm proposed by Price et al. [12]. They used the geodesic distance to measure the regional term. Based on the cost function in Eq. (3.4), the regional term is defined as

$$R_l(x_i) = s_l(x_i) + M_l(x_i) + G_l(x_i),\qquad(3.25)$$

where $s_l(x_i)$ is a term to represent user stokes, $M_l(x_i)$ is a global color model and $G_l(x_i)$ is the geodesic distance defined in Eq. (3.22). Mathematically, we have

$$M_F(x_i) = P_B(x_i), \quad M_B(x_i) = P_F(x_i),\qquad(3.26)$$

$$s_F(x_i) = \begin{cases} \infty, & \text{if } x_i \in \Omega_B, \\ 0, & \text{Otherwise}, \end{cases}\qquad(3.27)$$

and

$$s_B(x_i) = \begin{cases} \infty, & \text{if } x_i \in \Omega_F, \\ 0, & \text{Otherwise}. \end{cases}\qquad(3.28)$$

and

$$G_l(x_i) = \frac{D_l(x_i)}{D_F(x_i) + D_B(x_i)}, \quad l \in \{F, B\},\qquad(3.29)$$

which is normalized by the foreground and background geodesic distance $D_F(x_i)$ and $D_B(x_i)$.

Price et al. [12] redefined this regional term and minimized the cut cost in Eq. (3.4). The overall cost function becomes

$$E(L) = \lambda_R \cdot \sum_{i \in V} R_{L_i}(x_i) + \lambda_B \cdot \sum_{(i,j) \in E} E_2(i, j),\qquad(3.30)$$

where E_2 is the same boundary term as that in Eq. (3.16), and λ_R and λ_B are parameters used to weight the relative importance of the region and boundary components. A greater value in λ_R helps reduce the short-cutting problem, which is caused by a small boundary cost term. For robustness, Price et al. [12] introduced a global weighting parameter to control the estimation error of the color model and two local weighting parameters for the geodesic regional and boundary terms based on the local confidence of geodesic components.

The geodesic graph-cut outperforms the conventional graph-cut [6] and the geodesic segmentation [13]. It performs well when user interactions separate the foreground and background color distributions effectively as shown in Fig. 3.6.

3.1.6 Graph-Cut with Prior Constraints

Since the standard graph-cut method in Sect. 3.1.2 may fail in cases of objects with diffused or ambiguous boundaries, research has been done to mitigate the boundary problem by providing shape priors under the min-cut/max-flow optimization framework. A shape prior means the prior knowledge of a shape curve template provided by user interaction. Freedman et al. [15] introduced an energy term based on a shape prior by incorporating the distance between the segmented curve, c, and a template curve, $\bar{c}$, in the energy function as

$$E = (1 - \lambda)E_i + \lambda E_s \qquad (3.31)$$

where E_i is the energy function defined in Eq. (3.4), E_s is the energy term associated with the shape prior, and λ is a weight used to balance these two term. Specifically, E_s is in form of

$$E_s = \sum_{(i,j)\in E, l_{x_i} \neq l_{x_j}} \bar{\phi}(\frac{x_i + x_j}{2}), \qquad (3.32)$$

where x_i and x_j are neighboring pixels in image I, and l_x is the label of pixel x, and $\bar{\phi}(\cdot)$ is a distance function that all pixels x on the template curve $\bar{c}$ has $\bar{\phi}(x) = 0$. The final segmentation is obtained by minimizing the energy function in Eq. (3.31).

Veksler [16] implemented a star shape prior for convex object segmentation. As shown in Fig. 3.7, the star shape prior assumes that every single point x_j, which is on the straight line connecting the center, C_0, of the star shape and any point x_i inside the shape should also be inside the shape.

Veksler defined the shape constraint term as

$$E_s = \sum_{(x_i, x_j) \in \mathcal{N}} S_{x_i, x_j}(l_i, l_j), \qquad (3.33)$$

where

$$S_{x_i, x_j}(l_i, l_j) = \begin{cases} 0, & \text{if } l_i = l_j, \\ \infty, & \text{if } l_i = F \text{ and } l_j = B, \\ \beta, & \text{if } l_i = B \text{ and } l_j = F, \end{cases} \qquad (3.34)$$

Fig. 3.7 A *star shape* defined in [16]. Since *red point* x_i is inside the object, the *green point* x_j on the *line* connecting x_i with center C_0 should be labeled with the same label as x_i

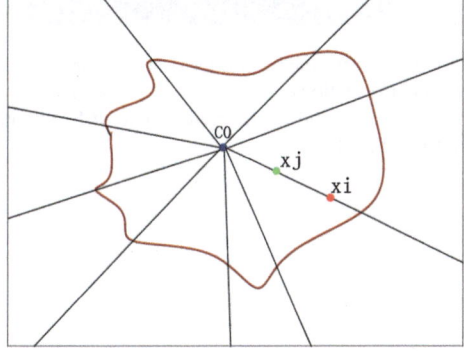

which is used to penalize the assignment of x_j with a label l_j different from that of x_i. Parameter β can be set as a negative value, which might encourage the long extension of the prior shape curve. The final segmentation is the optimal labeling obtained by minimizing the energy function in Eq. (3.31).

Being different from other interactive segmentation methods, a user just provides the center location of a star shape (rather than the strokes for foreground and background regions) in this system. The limitation of the star shape prior [16] is that it only works for star convex objects. To extend this star shape prior to objects of an arbitrary shape, Gulshan et al. [17] implemented multiple star constraints in the graph-cut optimization, and introduced the geodesic convexity to compute the distances from each pixel to the star center using the geodesic distance. We show how the shape constraint improves the result of object segmentation in Fig. 3.8.

Another drawback of the standard graph-cut method is that it tends to produce an incomplete segmentation on images with elongated thin objects. Vicente et al. [18] imposed a connectivity prior as a constraint. With additional marks for the disconnected pixels, their algorithm can modify the optimal object boundary so as to connect marked pixels/regions by calculating the Dijkstra graph cut [18]. This approach allows a user to explicitly specify whether a partition should be connected or disconnected to the main object region.

Fig. 3.8 Performance comparison of graph-cut segmentation with and without the shape constraint [17]. A flower is segmented out with a specified shape prior while other flowers are filtered out as background in the *right image*. **a** Segmentation by IGC [6]. **b** Segmentation with shape constraint [17]

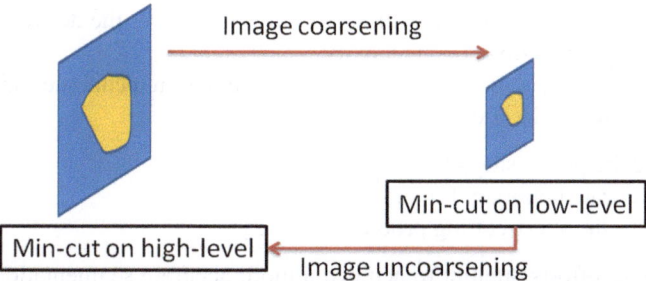

Fig. 3.9 Structure of multi-resolution graph-cut [20]. With the segmentation on low-level image, the computation of min-cut on high-level image is highly reduced

3.1.7 Multi-Resolution Graph-Cut

Besides improving the accuracy of the graph-cut segmentation, research has been conducted to increase segmentation efficiency, e.g., reducing the processing time and the memory requirement of a segmentation system (Fig. 3.9).

Wang et al. [19] proposed a pre-segmentation scheme based on the mean-shift algorithm to reduce the number of graph nodes in the optimization process, which will be detailed in Sect. 3.4.1.2. They also extended the approach to video segmentation, and proposed an additional spatiotemporal alpha matting scheme as a post-processing to refine the segmented boundary. To reduce the memory requirement for the processing of high resolution images, Lombaert et al. [20] proposed a scheme that conducts segmentation on a down-sampled input image, and then refines the segmentation result back to the original resolution level (Fig. 3.9). The complexity of the resulting algorithm can be near-linear, and the memory requirement is reduced while the segmentation quality can be preserved.

3.1.8 Discussion

The graph-cut segmentation method is popular in practical applications and becomes one of the most important interactive segmentation techniques because of its solid theoretical foundation and good performance. The min-cut/max-flow framework is based on the maximal a posteriori (MAP), which is the conditional probability of user interactions. The segmentation cost consists of a regional cost term, which is the posterior of the labeling with the knowledge of seed pixels labeled by the user, and a boundary term, which is used to locate object boundaries. A stress on either of the two terms will emphasize different aspects. With the global optimization, a graph-cut method can extract objects of interest with sufficient user interactions. To speed up, the global optimization process may not provide segmentation with pixel-wise accuracy in some cases. However, there are ways to recover the desired accuracy of object boundaries and object connectivity.

The improvement of the graph-cut based segmentation technique can be pursued along the following directions:

- increasing the processing speed [5, 20];
- finding accurate boundary [16, 17];
- overcoming the short-cutting problem [12, 13].

All of these efforts attempt to achieve a more accurate segmentation result at a faster speed with less user interaction.

3.2 Edge-Based Segmentation Methods

Edge detection techniques transform images into edge images by examining the changes in pixel amplitudes. Thus, one can extract meaningful object boundaries based on detected edges as well as prior knowledge from user interaction. In this section, we present edge-based segmentation methods and show how users can guide the process.

Edges, serving as basic features of an image, reveal the discontinuity of the image amplitude attribution or image texture properties. The location and strength of an edge provide important information of object boundaries and indicate the physical extent of objects in the image [21]. Edge detection refers to the process of identifying and locating sharp discontinuities in an image. It is the key and basic step toward image segmentation problems [22]. In the context of interactive segmentation, many algorithms have been proposed to segment objects of interest based on edge features, combined with user guidance and interaction. Live-wire and active contour are two basic methods that extract objects based on edge features. These two methods will be detailed after an overview on edge detectors.

3.2.1 Edge Detectors

Many edge-detection techniques based on different ideas and tools have been studied, including error minimization, objective function maximization, wavelet transform, morphology, genetic algorithms, neural networks, fuzzy logic, and the Bayesian approach. Among them, the differential-based edge detectors have the longest history, and they can be classified into two types: detection using the first-order derivative and the second-order derivative [22].

The first-order edge detectors calculate the first-order derivative at all pixels in an image. Examples include Sobel, Prewitt, Krisch, Robinson, and Frei-Chen operators. Sobel detectors are suitable for detecting edges along the horizontal and vertical directions while Roberts detectors work better for edges along 45° and 135° directions.

An operator involving only a small neighborhood is sensitive to noise in the image, which may result in inaccurate edge points. This problem can be alleviated by extending the neighborhood size [21]. The Canny edge detector was proposed in [23] to reduce the data amount while preserving the important structural information in an image. There have been a number of extensions of Canny's edge detector, e.g., [24–26].

The second-order edge detectors employ the spatial second-order differentiation to accentuate edges. The following two second-order derivative methods are popular:

- The Laplacian operator [27]
- The zero crossings of the Laplacian of an image indicate the presence of an edge. Furthermore, the edge direction can be determined during the zero-crossing detection process. The Laplacian of Gaussian (LoG) edge detector was proposed in [28] in which the Gaussian-shaped smoothing is performed before the application of the Laplacian operator.
- The directed second-order derivative operator [29]
- This detector first estimates the edge direction and, then, computes the one-dimensional second-order derivative along the edge direction [29].

For color edge detection, a color image contains not only the luminance information but also the chrominance information. Different color space can be used to represent the color information. A comparison of edge detection in RGB, YIQ, HSL and Lab space is given in [30]. Several definitions of a color edge have been examined in [31]. One is that an edge in a color image exists if and only if its luminance representation contains a monochrome edge. This definition ignores discontinuities in hue and saturation. Another one is to consider any of its constituent tristimulus components. A third one is to compute the sum of the magnitude (or the vector sum) of the gradients of all three color components.

3.2.2 Live-Wire Method and Intelligent Scissors

Live-wire boundary snapping for image segmentation was initially introduced in [32, 33]. This technique has been used in interactive segmentation in [34–41]. One of its implementations, called the Intelligent Scissors, has been widely used as an object selection tool in an image editing program, GIMP [42], and medical image segmentation applications. Intelligent Scissors can be well controlled even when the target image has a low contrast end weak edges.

Intelligent Scissors offer an object selection tool that allows rapid and accurate object segmentation from complex background using simple gesture motions with a mouse [32, 34, 35, 40, 41]. When a user sweeps the cursor around an object, a live-wire [32] automatically snaps to and wraps around detected object boundaries with real-time visual feedback. Since the user can control the mouse movement to guide the object boundary selection interactively, the segmentation result is generated according to user control and interaction.

The optimal object boundaries in Intelligent Scissors are obtained by imposing a weighted graph on the image and interactively computing the optimal path from a user selected seed point to all other possible path points using an efficient (linear-time) version of Dijkstra's graph search algorithm [43]. The process is detailed below.

The input image is viewed as a weighted graph, where nodes in the graph represent the pixel and directed and weighted edges are the links between each pixel with its 4-connected or 8-connected neighbors. The local cost of each directed link in this graph is the weighted sum of the component cost of image features such as Laplacian zero-crossing f_Z, gradient magnitude f_G and gradient direction f_D. The Laplacian zero-crossing is defined as

$$f_Z(i) = \begin{cases} 0 & \text{if } I_L(i) = 0 \\ 1 & \text{if } I_L(i) \neq 0 \end{cases} \tag{3.35}$$

where I_L is the Laplacian zero-crossing map of input image I and i is the node (or pixel) index. The gradient magnitude, f_G, is computed as an inverse linear ramp function [40] so that pixels of larger gradient magnitudes have smaller f_G. The gradient direction, f_D, is used to measure the directional consistency of each pixel with its neighbors.

The local cost $l(p, q)$ on the directed path from p to its neighboring pixel q is a weighted sum of component cost functions:

$$\begin{aligned} l(p, q) = w_Z \cdot f_Z(q) + w_G \cdot f_G(q) + w_D \cdot f_D(p, q) \\ + w_P \cdot f_P(q) + w_I \cdot f_I(q) + w_O \cdot f_O(q) \end{aligned} \tag{3.36}$$

where f_P, f_I and f_O denote the current, inside and outside values, respectively, which are defined as the pixel along, on the left and on the right of the boundary element [37]. Since f_Z, f_G and f_D are static cost functions, they can be computed initially. In contrast, f_P, f_I and f_O have to be updated dynamically since their values depend on the segmentation result. Note that pixels that have strong edge features will have a low local cost.

The shortest path cost from pixel p to seed point s, denoted by $c(p)$, is the minimum cumulative cost along the path from s to p. It can be calculated via

$$c(p) = \min\{c(q) + l(p, q)\}, \tag{3.37}$$

where q is a pixel in the neighbor of p, $c(q)$ is the shortest path cost from q to s, and $l(p, q)$ is given by Eq. (3.36).

A simple example of finding the shortest path from a seed with Dijkstra's algorithm [43] is shown in Fig. 3.10. For the ease of illustration, we use the static shortest path in this example.

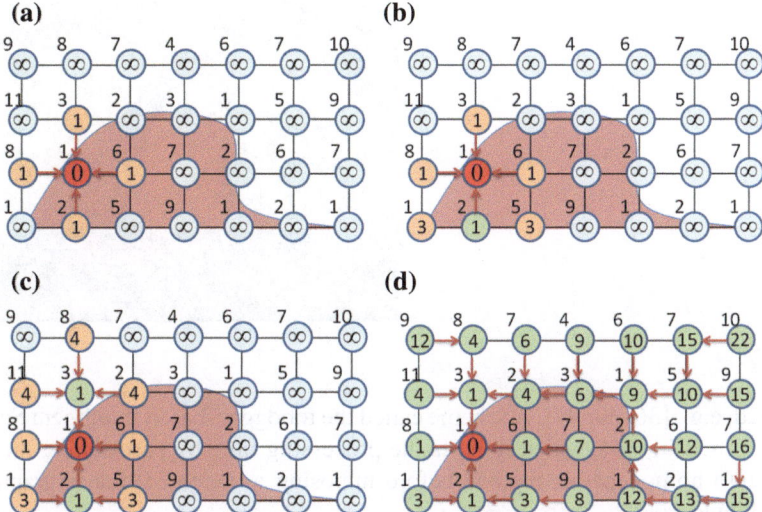

Fig. 3.10 An example of computing the shortest path cost from any node in the graph to the seed marked in *red* using Dijkstra's spanning tree algorithm [43]. **a** Initial static costs are computed and given on the *upper left* of each pixel. Each pixel is initialized with an infinite path cost to the seed point given inside the pixel. We start from the seed point in *red* and set its cost to zero. The cost of its neighbor node is the sum of the cost of the seed and the path cost to this seed, which is assumed to be 1. We add its neighbors to the node list $L = \{1, 1, 1, 1\}$ and sort the list. **b** Remove the head node from L, which is the southern node of the seed in this example. Add its path cost to its neighbors. If the new cost is smaller, update the path cost and the path direction. Add updated nodes to L, sort $L = \{1, 1, 1, 3, 3\}$. **c** Remove the head node from L, which is the northern node of the seed in this example. Add its path cost to its neighbors. If the new cost is smaller, update the path cost and the path direction. Add spanned nodes to L and sort L. **d** Iteratively span the graph until all nodes are spanned ($L = \Phi$). This is the spanning tree results and the shortest paths from all nodes in this graph to the seed are marked in *red*

When the cursor moves, this algorithm can compute an optimal path from the current position to the seed point automatically as shown in Fig. 3.10. Each optimal path is displayed, allowing the user to select an optimal object contour segment which visually corresponds to a portion of the desired object boundary. Figure 3.11 shows a segmentation result obtained by Intelligent Scissors.

Mortensen and Barrett [40] proposed "boundary cooling" and "on-the-fly training" to improve user experience. Boundary cooling relieves users of placing most seed points by automatically selecting a pixel on the current active live-wire segment that has a "stable" history to be a new seed point. The live-wire is restricted by "on-the-fly training," which refines the tracking boundary yet adheres to the specific type of current edge (rather than simply choosing the strongest edge).

Although a desired segmentation result can be obtained by involving a sufficient amount of human interaction and guidance, pre-calculation of the cost map for the graph in each tracking step often slows down the overall processing speed of Intelligent Scissors. To overcome this drawback, several acceleration strategies have been

Fig. 3.11 A segmentation example using Intelligent Scissors

proposed, e.g., [36–39, 44]. A scheme called the toboggan-based Intelligent Scissors was proposed in [44], which reduces the processing time by partitioning the input image into homogeneous regions before imposing a weighted planar graph onto region boundaries.

3.2.3 Active Contour Method

The active contour method is also called the snake or the deformable model. It is manually initialized to provide a rough approximation to an object boundary. Then, one can perform iterative contour refinement to determine the optimal boundary that minimizes an energy functional at each step. Being similar to Live-Wire, active contours use the edge features to derive an optimal object boundary under user guidance. All boundary points can be adjusted in parallel in an attempt to minimize the energy functional. This technique has been widely used in medical imaging when objects have similar shapes [45–47], where one can obtain prior knowledge of the desired contour conveniently.

Kass et al. [48] introduced a global minimum energy contour. The energy functional is a combination of internal forces (e.g., the boundary curvature and the distance between points) and external forces (e.g., the image gradient magnitude and the edge direction). In contrast with Intelligent Scissors which control the boundary seed points directly, the active contour method refines the shape of the initial contour via energy minimization. If the resulting boundary is not good, the process can be repeated with further boundary approximation. An active contour is globally optimal over the entire object contour space, while Intelligent Scissors compute the optimal path between a pixel and its closest seed point locally [35].

Given an initial approximation to a target contour, the active contour method locates the closest minimum energy contour by iteratively minimizing an energy functional. It combines internal forces to keep the active contour smooth and external forces to attract the snake to image features and constraint forces which help define the overall shape of the contour. The basic model is a controlled continuous spline under the constraint forces. The combined energy (or called the snake energy) of contour v can be written as

$$E_{snake}(v(s)) = \int_{s=0}^{1} \alpha(s)|\frac{dv}{ds}|^2 + \beta(s)|\frac{d^2v}{ds^2}|^2 - \gamma|\nabla I(v)|ds, \qquad (3.38)$$

where the first two terms are the internal energy terms used to measure the continuity and the smoothness of contour v, respectively, the last term is the external energy that takes into account how close v is to the actual contour in image I, and α, β and γ are weighting parameters. The contour evolves from the initial drawn contour to the desired object boundary by minimizing the energy functional iteratively.

3.2.4 Discussion

The edge-based interactive segmentation methods are designed to extract the object boundary out directly. Live-wire methods find the local shortest path defined on edge features so as to locate the object boundary. The active contour method applies both boundary and regional constraints to determine a well-shaped object contour. Both of them can be controlled by users to get better results. For an extensive study on contour detection techniques, we refer to a recent survey in [49], which presents an overview of edge- and line-oriented approaches to contour detection in the last two decades.

Since the original proposal of the active contour method [48], a lot of variants have been developed, falling broadly into three classes:

- Parametric active contour [48, 50–60];
- Non-parametric or geometric active contour [61–65];
- Physics inspired particle-based techniques [66, 67].

Liang et al. [68] proposed a "united snake" scheme that unifies several important snake varaints, such as the finite difference [48], the B-spline [50] and Hermite polynomial snakes [69] under the finite-element framework, and imposed the live-wire technique as a complementary hard constraint. As compared with edge-based snakes, region-based snakes [63, 70] are robust to image degradation and less sensitive to initialization since more global statistics are involved. Region-based snakes can be integrated with texture models for texture segmentation [71]. Several shape-based active contour methods, e.g., [72, 73], add an energy term, E_{shape}, to emphasize the shape force. In particular, the convex active contour (CAC) [73] performs well in locating object boundaries. We will compare the performance of various active contour methods in Chap. 4.

3.3 Random-Walk Methods

The random walk (RW) on a graph is a special case of the Markov Chain [74]. It has been wildly applied in computer vision, including image colorization, interactive image segmentation, automatic image segmentation and clustering, mesh segmentation and de-noising, shape presentation, image/stereo matching and image fusion. In interactive image segmentation, the RW algorithm offers a general-purpose multi-label (multi-object) image segmentation method that allows a user to initialize the background and object seeds [75].

As discussed before, image segmentation can be treated as a labeling problem that labels pixels as the foreground object or background, or as different objects in the multiple labeling context. The RW-based segmentation methods label an unseeded pixel by resolving the following question. Given a random walker starting at this location, what is the probability that it first reaches each of the seeded points [76]? The segmentation result is obtained by selecting the most probable seed destination of a random walker for each pixel.

To increase the processing speed, one can perform some offline task before the interactive segmentation procedure. For example, Grady et al. [77] proposed a scheme to compute several eigenvectors of the weighted Laplacian matrix of a graph and use this information to produce a linear-time approximation of the random walker segmentation algorithm. Kim et al. [78] used a generative model to segment images based on random walks with restart (RWR), which solves the weak boundary problem and the texture problem.

3.3.1 Random Walk (RW)

A simple example of 2-D RW on a graph is given in Fig. 3.12. In interactive image segmentation, a walker starts from a seeded node, s. At each step, this walker moves from its current position to one of its neighbors with a probability, which is specified

Fig. 3.12 An example of 2-D random walk on a graph

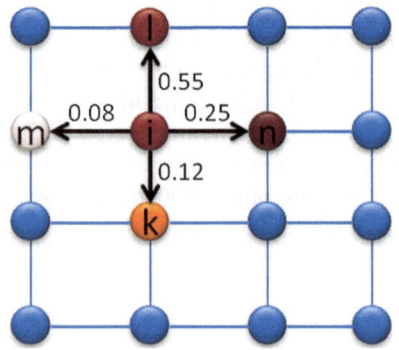

by the weight of the edge as shown in Fig. 3.12. In this example, the random walker at node i may move to nodes k, l, m and n in the next step with probabilities of 0.12, 0.55, 0.08 and 0.25, respectively.

Under the Markov chain assumption, the state of each node at a certain time instance is only related to the states of its neighbors. According to the Bayesian rule in Eq. (2.3), the posterior distribution of an image label is proportional to the product of the likelihood function and the prior distribution. When the prior label distribution of is uniform, the posterior label distribution is proportional to the label likelihood function. Then, the segmentation problem can be solved by finding the maximum likelihood. This is equivalent to the following problem: by starting from the pixel of interest, which seeded pixels s and t will this random walker reach first?

Given image $f(i)$, we can build an undirected graph $G = (V, E)$ as discussed in Sect. 3.1. The edge connecting vertices v_i and v_j is denoted by e_{ij}. The weight of edge e_{ij} is denoted by $w_{ij} = w(e_{ij})$, which is used to indicate a random walker bias from v_i to v_j. In the undirected graph, we have $w_{ij} = w_{ji}$, and $w_{ij} > 0$. The degree of a vertex v_i is $d_i = \sum w(e_{ij})$ for all edges e_{ij} incident on v_i.

Let $\mathbf{P}$ be the transition matrix [74] of an RW. The probability of a random walker to stay at node i after t iterations can be written as

$$\pi^{(t)} = \mathbf{P}\pi^{(t-1)} = \mathbf{P}^t \pi^{(0)}. \tag{3.39}$$

If the steady-state vector π exists, we can obtain

$$\mathbf{P}\pi = \pi. \tag{3.40}$$

We see from above that π is an eigenvector of $\mathbf{P}$ with eigenvalue equal to 1. As stated in [76], the random walker probabilities have the same solutions as the combinatorial Dirichlet problem. The Dirichlet integral was defined as [79]:

$$D[\mu] = \frac{1}{2} \int_\Omega |\nabla \mu|^2 d\Omega \tag{3.41}$$

over a field $\mu \in \Omega$. The harmonic function satisfying the Laplace equation is

$$\nabla^2 \mu = 0. \tag{3.42}$$

The harmonic function that satisfies the boundary conditions can minimize the Dirichlet integral in Eq. (3.41), since the Laplace equation is the Euler-Lagrange equation for the Dirichlet integral [79]. The combinatorial Laplacian matrix can be defined as [80]:

$$L_{ij} = \begin{cases} d_i, & \text{when } i = j, \\ -w_{ij}, & \text{when } v_i \text{ and } v_j \text{ are connected,} \\ 0, & \text{otherwise.} \end{cases} \tag{3.43}$$

where $d_i = \sum_j w_{ij}$. The $m \times n$ edge-node incidence matrix is defined as

$$A_{e_{ij}v_k} = \begin{cases} 1 & \text{when} \quad i = k \\ -1 & \text{when} \quad j = k \\ 0 & \text{otherwise} \end{cases} \tag{3.44}$$

A can be interpreted as a combinatorial gradient operator while A^T as a combinatorial divergence. Matrix L can be decomposed into $L = A^T C A$ [76], where C is an $m \times m$ diagonal matrix with the edge weights along the diagonal. The Dirichlet integral in Eq. (3.41) can be approximated via

$$D[x] = x^T L x = \frac{1}{2}(Ax)^T C(Ax) = \frac{1}{2}x^T L x = \frac{1}{2} \sum_{e_{ij} \in E} w_{ij}(x_i - x_j)^2. \tag{3.45}$$

Since L is positive semidefinite, the critical points of $D[x]$ are the minima and the combinatorial harmonic is the vector, x, that minimizes Eq. (3.45).

The seeded (or marked) vertices and the unseeded (or unknown) vertices in graph G are denoted by V_M as V_U, respectively. We have

$$V_M \bigcup V_U = V, \text{ and } V_M \bigcap V_U = \phi. \tag{3.46}$$

We order L and x to ensure seeded vertices come first and unseeded points next. Then, Eq. (3.45) can written as

$$\begin{aligned} D[x] &= \frac{1}{2}\left| x_M^T \, x_U^T \right| \left| \begin{matrix} L_M & B \\ B^T & L_U \end{matrix} \right| \left| \begin{matrix} x_M \\ x_U \end{matrix} \right| \\ &= \frac{1}{2}(x_M^T L_M x_M + 2x_U^T B^T x_M + x_U^T L_U x_U) \end{aligned} \tag{3.47}$$

where x_M and x_U correspond to the potentials of the seeded and the unseeded vertices. To find the critical point, we differentiate $D[x]$ with respect to x_U and set it to zero. Then, we have

$$L_U x_U = -B^T x_M, \tag{3.48}$$

which is a system of linear equations with V_U unknowns. Random walkers in G connect each unknown vertex with a seed node. Then, for each label, s, the solution to the combinatorial Dirichlet problem can be found by solving

$$L_U x^s = -B^T m^s, \tag{3.49}$$

where x_i^s is the probability of labeling vertex v_i as s and $m_i^s = 1$ when vertex v_i is labeled with s_i; otherwise, $m_i^s = 0$. Equation (3.49) can also be written with respect to all labels $s = 1, \ldots, K$, where K is the total number of labels, in form of

$$L_U X = -B^T M \tag{3.50}$$

where both X and M have K columns with each column for one seed. For each node, the probabilities of labeling sum to unity:

$$\sum_s x_i^s = 1, \quad \forall v_i \in V. \tag{3.51}$$

There are only $K - 1$ sparse linear systems to solve. For binary segmentation with $K = 2$, there is only one sparse linear system left.

To summarize, the RW segmentation algorithm can be stated as follows:

1. The image intensities are mapped to edge weights via

$$w(e_{ij}) = \exp(-\beta(g_i - g_j)^2), \tag{3.52}$$

   where g_i is the image intensity at pixel i, β is a parameter to normalize the square gradients $(g_i - g_j)^2$.
2. Based on user's interaction, seeded nodes for a set, V_M. The probability at an unknown node can be found by solving Eq. (3.49), or the probabilities of all unknown nodes can be determined at once by solving Eq. (3.50).
3. To determine the most probable seeded node, we select $max_s(x_i^s)$ for vertices v_i, and assign the label of $max_s(x_i^s)$ to v_i.

Then, the final segmentation is obtained.

Grady et al. [76] demonstrated that the RW method yields a better segmentation result than the graph-cut method [1] in terms of robustness to noise, weak boundary detection and ambiguous region segmentation. An example is given in Figs. 3.13 and 3.14.

Recently, a constrained RW algorithm was proposed in [81], which considers three types of user interaction (namely, the foreground and background seed input,

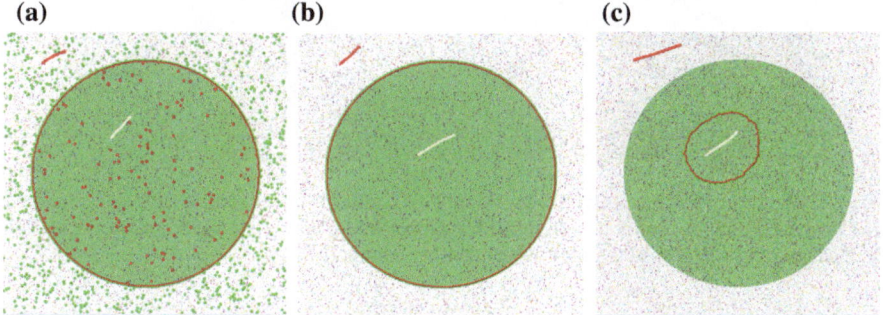

Fig. 3.13 Robustness of the RW segmentation when applied to a noisy image [76]. When the noise is too strong, RW fails to segment the object in (**c**). **a** Noisy image segmentation by IGC [1]. **b** Noisy image segmentation by RW [76]. **c** More noisy image segmentation by RW [76]

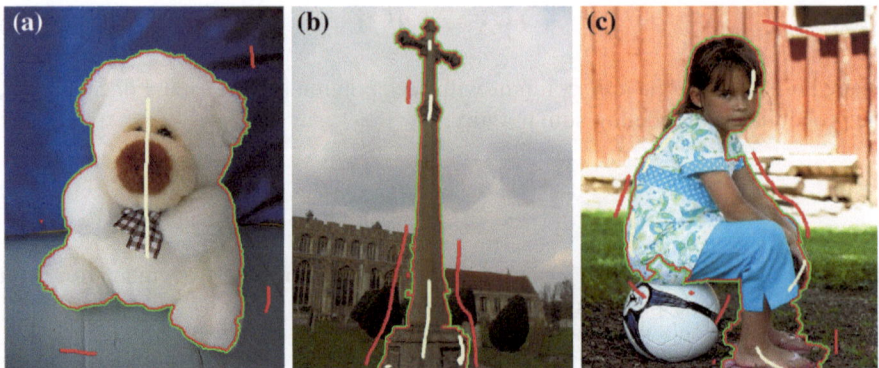

Fig. 3.14   Segmentation results by RW [76]

the soft constraint input and the hard constraint input) as well as their combinations. The soft constraint input allows a user to indicate whether there is a boundary passing through this region. Being assisted with the soft input, the segmentation result can be easily guided as shown in Fig. 3.15.

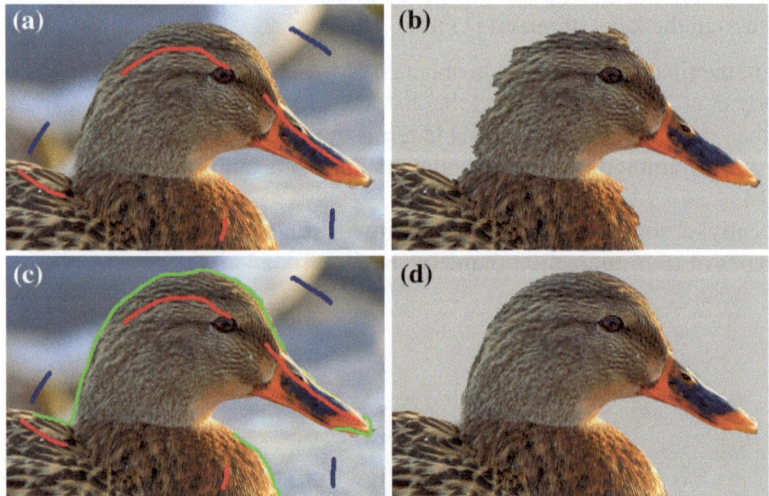

Fig. 3.15   Segmentation results with the soft input [81]. **a** Hard strokes for RW: *red* for foreground; *blue* for background. **b** Segmentation result with input (**a**). **c** Soft input: *green strokes* to indicate the location of boundaries. **d** Segmentation result with input (**c**)

3.3.2 Random Walk with Restart (RWR)

One problem of RW is that a different initialization of seeded points may still produce the same steady-state vector, π. Kim et al. [78] proposed a generative image segmentation scheme by using random walk with restart (RWR). To enhance the importance of the initialization process, RWR constrains a random walker by setting up a restarting probability. That is, at each step of RWR, the walker has two choices: randomly moving to one of it neighbors with probability c and jump back to its initial seed point and restart its random walking with probability $1 - c$.

Consider a joint distribution model of pixels and labels, $p(l_k, x_i)$, where $x_i \in X = \{x_1, x_2, \ldots, x_N\}$ is a pixel, and $l_k \in L = \{l_1, l_2, \ldots, l_K\}$ is a label that takes K different values. The joint distribution can be computed by the prior label probability $p(l_k)$ and the pixel likelihood $p(x_i|l_k)$ of the RWR process. By the Bayes' rule, the posterior probability of the RWR process can be written as

$$p(l_k|x_i) = \frac{p(x_i|l_k)p(l_k)}{\sum_{n=1}^{K} p(x|l_n)p(l_n)}. \tag{3.53}$$

Assume that there are more than one seed points for each label, and $S^{l_k} = \{s_1^{l_k}, s_2^{l_k}, \ldots, s_{M_k}^{l_k}\}$ is the set of M_k seed pixels with label l_k. Then, we can represent the likelihood as

$$p(x_i|l_k) = \frac{1}{z} \sum_{m=1}^{M_k} p(x_i|s_m^{l_k}, l_k)p(s_m^{l_k}|l_k) = \frac{1}{z \times M_k} \sum_{m=1}^{M_k} p(x_i|s_m^{l_k}, l_k), \tag{3.54}$$

where Z is a normalization constant. Each likelihood of pixel x_i is modeled by a mixture of distribution $p(x_i|s_m^{l_k}, l_k)$ from each seed $s_m^{l_k}$, which has a seed distribution $p(s_m^{l_k}|l_k)$.

Under the assumption that the seed distribution is uniform (i.e., $1/M_k$), the likelihood of RWR in Eq. (3.54) will be calculated as the average of the pixel distributions of all seed pixels with label l_k. This makes RWR robust against the number of seed pixels.

The overall RWR segmentation algorithm can be described below.

- Build a weighted graph for an image. In [78], the weight between pixels x_i and x_j is defined as the typical Gaussian weighting function given by

$$w_{ij} = \exp(-\frac{||g_i - g_j||^2}{\sigma}). \tag{3.55}$$

- Calculate the likelihood in Eq. (3.54) using the RWR process and, then, assign the label with the maximum posterior probability in Eq. (3.53) to each pixel in the graph.

The link between the RWR process in graph G and the computation of Eq. (3.54) will be explained below.

Consider a random walker who starts from the m-th seed with label l_k (or $x_i = s_m^{l_k}$) with a restarting probability c. The steady-state probability for this random walker to stay at pixel x_i, can be written as

$$p(x_i | s_m^{l_k}, l_k) \approx r_{im}^{l_k} \tag{3.56}$$

For N nodes on a graph, we can form the N-dimensional vector $r_m^{l_k} = [r_{im}^{l_k}]_{N \times 1}$. The adjacent matrix for all edges in graph G is $W = [w_{ij}]_{N \times N}$, where w_{ij} is defined in Eq. (3.55). Then, we have the following relationship [82]:

$$
\begin{aligned}
r_m^{l_k} &= (1 - c) P r_m^{l_k} + c b_m^{l_k} \\
&= c(I - (1 - c)P)^{-1} b_m^{l_k} \\
&= Q b_m^{l_k}
\end{aligned}
\tag{3.57}
$$

where $b_m^{l_k} = [b_i]_{N \times 1}$ is the N-dimensional vector with a binary value ($b_i = 1$ when $x_i = s_m^{l_k}$, and $b_i = 0$ otherwise), $Q = [q_{ij}]_{N \times N}$ is a matrix to compute the affinity between two pixels, and $P = [p_{ij}]_{N \times N}$ is the transition matrix. It is easy to show that P is the row-normalized adjacency matrix W; namely, $P = D^{-1}W$, where $D = diag(D_1, D_2, \cdots, D_N)$ and where $D_i = \sum_{j=1}^{N} w_{ij}$. Then, through Eq. (3.54), we have

$$[p(x_i | l_k)]_{N \times 1} = \frac{1}{Z \times M_k} Q b^{l_k}. \tag{3.58}$$

For Q, q_{ij} is the likelihood that x_i has the same label as x_j. Thus, we have

$$Q = c(I - (1 - c)P)^{-1} = c \sum_{t=0}^{\infty} (1 - c)^t P^t, \tag{3.59}$$

where P^t is the t-th order transition matrix, p_{ij}^t is the total probability for a random walker starting from x_i walks to x_j after t iterations, considering all possible paths between these two pixels. As t increases, we iteratively update P^t until it converges. The resulting matrix Q can be solved via matrix inversion as shown in Eq. (3.59). The label s_i of pixel x_i is decided by

$$s_i = \arg \max_{l_k} p(l_k | x_i) = \arg \max_{l_k} p(x_i | l_k). \tag{3.60}$$

By assigning a label to each pixel in the graph, we obtain the final segmentation.

As compared with the graph-cut method [1] and the random walker method [76], the RWR algorithm can segment images with weak edges and textures more efficiently. Examples are shown in Figs. 3.16 and 3.17.

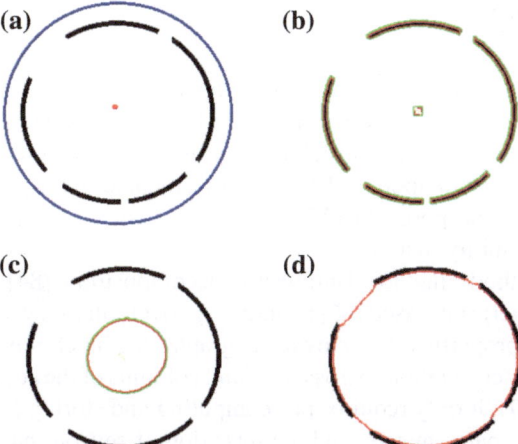

Fig. 3.16 Performance comparison between the graph-cut (GC), the random-walker (RW) and the random walker with restart (RWR) for an image with weak edges. **a** Test image with user stokes: *red* for foreground; *blue* for background. **b** Segmentation result by IGC [1]. **c** Segmentation result by RW [76]. **d** Segmentation result by RWR [78].

Fig. 3.17 Performance comparison between the graph-cut (GC), the random-walker (RW) and the random walker with restart (RWR) for an textured image. **a** Test image with user stokes: *white* for foreground; *red* for background. **b** Segmentation result by IGC [1]. **c** Segmentation result by RW [76]. **d** Segmentation result by RWR [78].

3.3.3 Discussion

The exact solution of both RW and RWR demands matrix inversion, where the matrix is often diagonally concentrated and of a large dimension [83]. How to reduce the computational complexity to facilitate online segmentation applications is an interesting problem. Built upon the RW image segmentation [76], a fast approximate random walker was proposed in [77] by pre-computing an approximation to the random walk probability matrix.

Fast RWR methods include Hub-vector decomposition [84], block-structure-based [85] and fingerprint-based [86] methods. A novel solution was proposed in [83] by exploiting two properties shared by many graphs: the block-wise community-like structure and linear correlations across rows and columns of the adjacency matrix. As compared with [77], it only requires pre-computing and storing a low-rank approximation to a large sparse matrix, and the inversion of several matrices of a smaller size.

3.4 Region-Based Methods

Region-based interactive image segmentation methods segment images based on regions rather than pixels. They partition an input image into regions that are similar according to a set of criteria. Thresholding [87], region grow [88], region splitting and region merging [89] are main examples in this category.

3.4.1 Pre-Processing for Region-Based Segmentation

Before elaborating on region-based interactive segmentation methods, we have a brief review on the pre-processing step for region-based segmentation in this subsection. Two well known algorithms are the watershed and the mean-shift algorithms, which are developed for the processing of monochrome and color images, respectively.

3.4.1.1 Watershed Algorithm

The watershed algorithm, also called the watershed transform, borrows tools from mathematical morphology [90]. Topographic and hydrology concepts are useful in the development of region segmentation methods. In this context, a monochrome image is regarded as an altitude surface in which high-amplitude pixels correspond to ridge points while low-amplitude pixels correspond to valley points. If a drop of water falls on a point of the altitude surface, it moves to a lower altitude until it reaches a local altitude minimum. The accumulation of water in the vicinity of a local minimum is called a catchment basin. All points that drain into a common catchment basin are part of the same watershed. A valley is a region that is surrounded by a ridge. A ridge is the location of the maximum gradient of the altitude surface.

There are two approaches to watershed computation in an image [21, 87, 91], as described below.

- The rainfall approach
- The local minima in an image are first located. Each local minimum is given a unique tag. Adjacent local minima are combined with a unique tag. When a conceptual water drop is placed at each untagged pixel, it moves to its lower-amplitude neighbor until the drop reaches a tagged pixel. Then, this pixel is assigned with this tag value.
- The flooding approach
- A conceptual single pixel hole is pierced at every local minimum. The amplitude surface is lowered into a large body of water. The water enters the holes and proceeds to fill each catchment basin. If a basin is about to overflow, a conceptual dam is built on its surrounding ridge line to a height equal to the highest altitude ridge point.

The watershed algorithm tends to produce over-segmented results as shown in Fig. 3.18. Research has been done to alleviate this over-segmentation effect [92].

3.4.1.2 Mean-Shift Algorithm

The mean-shift algorithm is a non-parametric clustering technique for the analysis of a complex multi-modal feature space and for delineating arbitrarily shaped clusters. It searches for local maximal density points and groups pixels into clusters defined by these maximal density points.

The rationale behind this clustering approach is that the feature space can be regarded as an empirical probability density function (PDF) of the represented feature

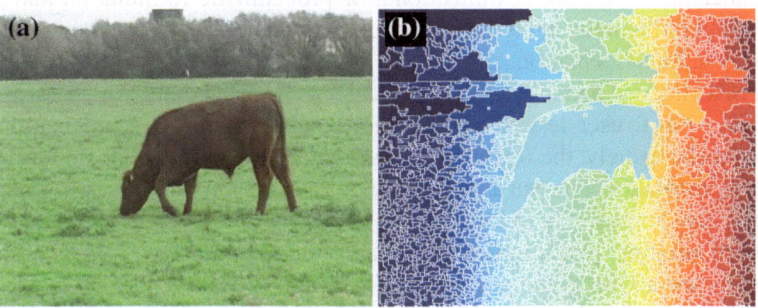

Fig. 3.18 An watershed example, which yields over-segmented subregions. **a** A cow image. **b** Its watersheding result with colored subregions

Fig. 3.19 A mean-shift segmentation example from [93], which yields an over-segmented result. **a** A butterfly image. **b** Its mean-shift segmentation result with subregion boundaries colored in *white*

vector. Dense regions in the feature space correspond to local maximum of the PDF, which offers one mode of the unknown density. Once the location of the mode is determined, the cluster associated with it is delineated based on the local structure of the feature space.

Figure 3.19 gives an example of the meanshift segmentation result from [93], where homogeneous pixels are clustered in one subregion. It offers an over-segmented result.

3.4.2 Seeded Region Growing (SRG)

The seeded region growing (SRG) method was proposed by Adams and Bischof [88] for interactive image segmentation. Although it does not have a solid mathematical basis under the statistical, optimizational or probabilistic formulation and suffers from certain limitations, it has gained popularity due to its fast processing speed and simplicity in implementation.

The SRG method demands a user to select a set of seed points, which are assigned labels according to user desired segmentation result. For a binary segmentation, there are two labels; namely, the foreground label and the background label. For multiple objects segmentation, one can assign a label to each object. Two binary segmentation examples are shown in Fig. 3.20.

At each iteration, pixels in subregions adjacent to the foreground or background subregions are added to the active set. The initial seeds are then replaced by the centroids of generated homogeneous regions by involving added pixels. All pixels in the new subregion are assigned to the label of the seed point in the same subregion. If a pixel in a subregion encounters two or more labels, it chooses the one with the

Fig. 3.20 Two segmentation examples by using the SRG method [3]. **a** A flower image with simple markups, where *red* for the foreground object and *blue* for background. **b** The segmentation result of the flower image, where the flower is segmented out completely. **c** A cat image with simple markups, where *red* for the foreground object and *blue* for background. **d** The segmentation result of the cat image, where part of background is confused to be the foreground, and vice versa

minimum distance to the average color of pixels in the subregion. The algorithm iterates until all pixels in the image are assigned to a label. The segmentation results can be obtained by extracting pixels with the same label.

Figure 3.20 shows two segmentation examples of the SRG method from [3]. We see that the SRG method produces a good result for the flower image, where the object has sharp boundaries and its color is distinguished from the background. However, it does not produce a satisfactory result for the cat image, part of background is confused to be the foreground, and vice versa, due to the color similarity of the foreground and background.

The SRG method suffers from the pixel sorting problem in propagating seed points. Mehnert and Jackway [94] showed that a different order of processing pixels leads to different final segmentation results. Since region boundaries are used to define the boundaries of different image components, Fan et al. [95] proposed a boundary-oriented technique to accelerate the seeded pixel labeling procedure as well as an automatic seed point selection method.

3.4.3 GrowCut

The GrowCut method [96] is a foreground extraction and background removal tool. It is implemented as a plug-in for Photoshop-compatible graphics editors. Being similar the graph-cut methods, the GrowCut method let user provide seed points to indicate the objects of interest. GrowCut allows user correction and guidance in the segmentation process, providing more efficient control and, thus, leading to more accurate segmentation results especially in some challenging cases [96].

Although GrowCut has a name similar to Graph cut and GrabCut, it is actually quite different. It is a region growing method with its root in Cellular automata (CA). GrowCut was first proposed by Von Neumann and Burks in [97]. Its theory and applications were recently reviewed in [98, 99]. Its simplicity and ease of parallel implementation make it a popular choice in the area of medical image processing where the data dimension could be 3D or 4D (space plus time) [100].

A bi-direction deterministic CA is used in GrowCut to propagate the label of one pixel to its neighborhood based on the color similarity and the label strength. An N-dimensional ($N \geq 2$) image is presented as a triplet, $A = (S, N, \delta)$, where S is the set of label states of pixels, N is the neighborhood of pixels (called the Neumann neighborhood or the Moore neighborhood) [99], and δ is the rule to propagate the labels of seed points and update the state of each pixel during the iteration. In each iteration, the pixel state is updated according to rule δ by considering the pixel strength and its color dissimilarity with their neighbors [96]. The final propagation result is obtained when the iterated propagation process converges.

To smoothen the segmented boundaries of an object, Vezhnevets et al. [96] defined an additional rule to control the propagation of states along boundaries. Since the strength of pixel states can be enhanced by user input, the interaction can serve as a soft constraint and guide the system to reach the desired segmentation result while reducing the error in the region controlled by user input.

The GrowCut method can be applied to multiple object segmentation at once, since the states in the iteration can have multiple values. Each state represents the label of one object. Since the label propagation is performed locally and independently, the GrowCut method allows parallel implementation for speed-up.

The accuracy of GrowCut's segmentation is highly dependent on image content and user interaction. The performance of the Growcut method, the Interactive Graph Cut (IGC) method [1], the Grabcut method [4] and the Random Walk method [101] was compared by Vezhnevets et al. in [96]. For the GrowCut plug-in for Photoshop, please refer to (http://growcut.com).

3.4.4 Maximal Similarity-Based Region Merging

The Maximal Similarity-based Region Merging (MSRM) method [102] begins with the over-segmentation result of the mean shift algorithm. User strokes are used to indicate the position and main features of the object and background. The non-markup background subregions are automatically merged and labeled, while the non-markup foreground subregions are identified and prevented from being merged with background. Once all non-markup subregions are labeled, the foreground object contour can be readily extracted.

The mean-shift algorithm as a pre-processing step can be replaced by other algorithms such as the watershed algorithm and the level set algorithm. The region merging process is conducted by merging a region into one of its neighboring regions, which has the most similarity. A region can be described in several attributes such as color, edge, texture, shape, and size. The color histogram was chosen in [102] as an effective region descriptor.

To measure the similarity between regions R and Q, one criterion is

$$\rho(R, Q) = \sum_{\mu=1}^{4096} \sqrt{Hist_R^{\mu} \cdot Hist_Q^{\mu}}, \qquad (3.61)$$

where $Hist_R$ and $Hist_Q$ are the normalized histogram of R and Q, in which each color channel is quantized into 16 levels, and the histogram of each region is calculated in the RGB space of $16 \times 16 \times 16 = 4096$ bins. The higher the $\rho(R, Q)$ is, the higher the similarity between them. Although two perceptually different regions may have similar histograms, such a likelihood is very low for neighboring regions in real-world images.

As shown in Fig. 3.21, the MSRM process consists of two stages, which are repeatedly executed until no new merging occurs. Its strategy is to merge as many background regions as possible while keeping the foreground region from being merged. Thus, its first stage is to start from background regions. For each neighboring region of a background region, MSRM calculates its similarities with its neighbor regions. If the most similar neighbored region has been labeled as background, MSRM merges this region into background. Whenever there is a merging process, the histogram of the new merged region is updated. Because of the histogram statistics and similarity calculation in each merging step, the overall computational complexity is high. This background merging procedure is iteratively implemented until, for all the background regions, there is no region to be merged.

After the background merging stage, there are still non-marker regions left, which might be foreground or background. For each non-marker region, one can calculate the similarities between it and its adjacent regions. Then, one can assign the same label to this region with its adjacent region, which has the largest similarity. This procedure is iteratively performed until all regions are labeled.

The MSRM method can extract user's interested object with a sharper contour than the Interactive Graph Cut (IGC) method [6]. On the other hand, it has some limitations. Accurate extraction requires sufficient user input to cover the main feature regions. Being similar to [103], the MSRM method relies on initial over-segmentation results. If the initial segmentation does not provide a good basis for region merging,

Fig. 3.21 A segmentation example by using the MSRM method [102]. **a** An initial over-segmented image with user's markups, where *green* is for the object and *blue* for background. **b** An intermediate result to show the background merging process. **c** Further background merging result. **d** All background regions are merged. **e** The remaining non-markup regions are merged

the MSRM method may fail. On the other hand, if there are too many over-segmented regions, the MSRM method will demand longer merging time.

3.4.5 Region-Based Graph Matching

The matching attributed relational graph (MARG) method [103] also begins with over-segmented sub-regions. User labeled strokes are propagated by exploiting the color and structure information on the graph constructed by over-segmented subregions. The objects of interest from the input image are indicated by user interaction such as scribbles drawn over the image. The main region merging procedure is performed by matching two graphs: the input graph, which represents the over-segmented image; and the model graph, which is constructed by sub-regions that have user-assigned labels.

Noma et al. [103] applied the watershed algorithm [87] and generated an initial segmentation of the input image with many small sub-regions, each of which is homogeneous in attributes. Each sub-region created by this over-segmentation procedure is described as a graph node. The entire input image is presented as an attributed relational graph (ARG), $G(V, E, u, v)$, which can represent the image color information and the structural information (spatial relations) via user scribbles. An ARG contains the following:

- V is a set of vertices, each of which represents the centroid of a sub-region;
- E is a set of edges connecting graph nodes;
- Attribute u is the property of each sub-region;
- Attribute v is the property of each edge.

Attribute u is typically chosen to be the normalized mean intensity of a sub-region. Any two adjacent sub-regions have an edge. To explore the structural information of the graph, each directed edge is assigned a vector denoted by v, which has a direction and a magnitude. The relationship of edge vectors v_1 and v_2 can be characterized by [104]:

$$C_{vec}(v_1, v_2) = \lambda_2 \frac{|\cos\theta - 1|}{2} + (1 - \lambda_2)\frac{||v_1| - |v_2||}{C_S},\qquad(3.62)$$

where θ is the angle between vectors v_1 and v_2 and $|v_1|$ and $|v_2|$ are their magnitudes.

The graph of an input image yields the input graph, G_i, while the image with user scribbles can be represented by a model graph denoted by G_m. One can propagate user labels by conducting a mapping between these two ARGs. As there are different node numbers in the two ARGs, this mapping is an inexact homomorphism [104, 105].

Noma et al. [106] proposed a deformed graph approach to match G_i and G_m. A deformed graph is constructed from the nodes of the input graph G_i. The process of deformed graph matching is showed in Fig. 3.22. The deformed graph matching is calculated by minimizing the following cost function:

$$E(v_i, v_m) = \lambda_1 d_A + (1 - \lambda_1) \sum_{\text{deformed edges}} d_S,\qquad(3.63)$$

where d_A evaluates the dissimilarity between the attribution of the deformed vertex, v_d, and the model vertex, v_m, in form of

$$d_A(v_d, v_m) \propto d(u(v_d), u(v_m)),\qquad(3.64)$$

where $d(\cdot, \cdot)$ is the Euclidean distance. The cost function in Eq. (3.63) attempts to balance the weight distributions of the graph intensity term, d_A, and the structure term, d_S. The structure term of matching G_d and G_m is defined as

$$d_S(G_d(v_i, v_m), G_m) = \frac{1}{|E(v_d)|} \sum_{e_d \in E(v_d)} C_{vec}(v(e_d), v(e_m)), \qquad (3.65)$$

where $G_d(v_i, v_m)$ denotes the deformed graph by taking v_i as v_d. Thus, the matching of two graph is only related to the deformed vertex and its connected vertices, since other vertices and edges remain the same as illustrated in Fig. 3.22.

By finding the minimum energy cost of matching, each v_i of G_i can be matched to a vertex, v_m, of model graph G_m and assigned with the same label of its matched vertex v_m. The segmentation result can be generated by extracting the partition of the image with the same label (Fig. 3.23).

The MARG method can be easily extended to multiple object segmentation, since all initialized vertex sets with user labels can get their matches by the same matching procedure. To get better results, two post-processing strategies were proposed in [103] to fill in holes in objects and remove isolated partitions. In [5], a simple pixel-based refinement was applied along the object boundary to enhance robustness and accuracy. Some segmentation results are shown in Fig. 3.24.

Since MARG handles the structure similarity of an object based on user scribbles, Noma et al. [106] attempted to segment multiple images of a similar structure (i.e., images from a sequence) at once. They built model graph G_m based on one image with user scribbles and, then, used this model graph to segment other images containing the same object with a changing location.

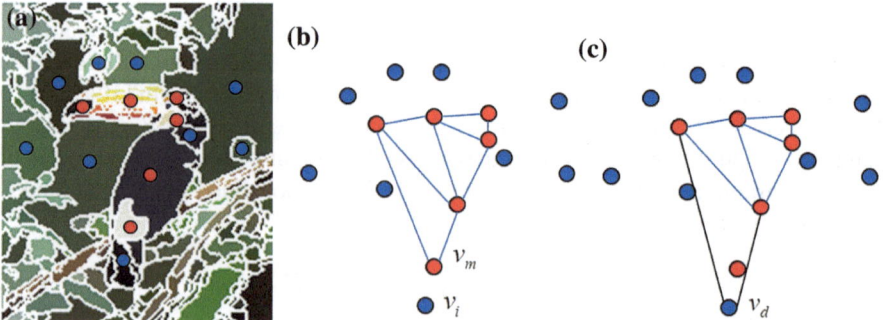

Fig. 3.22 Illustration of the deformed graph matching [103]. **a** Some of the graph nodes from the over-segmented image. **b** Model graph with user labeled nodes v_m (in *red*). **c** Deformed graph with node v_d

3.4.6 Discussion

The segmentation methods based on region merging/splitting and growing provide an efficient way to extract object regions. The SRG and the GrowCut methods are region growing methods starting from the label of seed pixels. The can be computed quickly due to its simplicity. The MSRM and MARG methods start from an initial over-segmented image. Although the computation within a region can be saved, the final segmentation quality will be affected by the initial segmentation. For example, when there is a weak edge that cannot be well detected in the over-segmentation process, the edge will not be detected in the region merging/growing process. To handle this problem, a boundary refinement procedure has to be applied to re-classify boundary pixels.

3.5 Local Boundary Refinement

In the last several sections, we introduced methods that either track object boundary or propagate user labels throughout the input image. An object can be extracted out by a looped contour or with the same label. In this section, we examine two post-processing techniques to improve the accuracy of segmentation results.

One post-processing technique is to merge a small isolated sub-region to its surrounding sub-region as proposed in [103]. For example, as shown in Fig. 3.25, region A is a small region without a label, it can be merged into its surrounding region by this post-processing technique. On the other hand, in order to segment this small region out, a user should assign a label to this region explicitly.

For more accurate segmentation of the object boundary, we can develop a pixel-based refinement scheme along the boundary [4, 5]. In Lazy Snapping, Li et al. [5] represented an object boundary using triangles and allowed users to edit the boundary

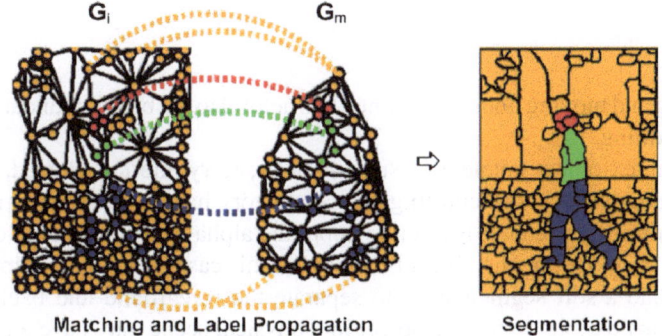

G_i G_m

Matching and Label Propagation   Segmentation

Fig. 3.23 Label propagation in the MARG method [103]

Fig. 3.24 **a** Input images with scribbles; **b** segmented masks from MARG; and **c** composition with new background [103]

Fig. 3.25 An isolated region *A* without a label can be merged into its surrounding partition [103]

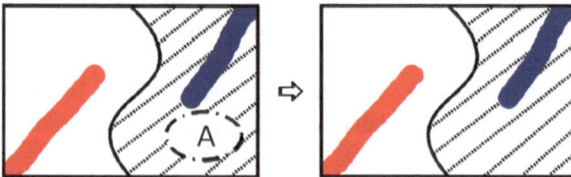

by dragging and moving boundary points. It is helpful to get better boundary locations with user editing.

For complex object boundaries such as hairy, furry, motion blurred, and transparent boundaries, it is difficult to get a satisfactory hard-segmentation result even with user editing. Instead, it is often to apply the alpha matting technique as a post-processing step to refine the object boundary in this case. The goal of alpha matting is to calculate a soft segmentation to separate the foreground and background as accurately as possible. Since the object in a physical scene may have a finer spatial resolution than the size of a discretized image pixel, one pixel may contain a mix of both the foreground and background information [107].

The soft segmentation can be represented in the form of

$$I(x, y) = \alpha(x, y)F(x, y) + [1 - \alpha(x, y)]B(x, y), \qquad (3.66)$$

where $I(x, y)$ is the observed image value at pixel (x, y), $F(x, y)$ are $B(x, y)$ are the foreground and background values at (x, y), and $0 \leq \alpha(x, y) \leq 1$ is the alpha matte function. This model was first proposed in [108] for the purpose of anti-aliasing in image segmentation. Each pixel along the object boundary is the blending result of foreground and background colors on the boundary, and the alpha value controls the weight of the foreground color.

When we calculate Eq. (3.66) in the color space, there are 7 unknowns (namely, colors of $F(x, y)$ and $B(x, y)$ and the alpha value $\alpha(x, y)$) to be determined with only known value $I(x, y)$ at each pixel location. Thus, the problem is ill-posed. Many regularization schemes have been proposed for Eq. (3.66) by setting constraints on F, B and α [13, 107, 109, 110]. Generally speaking, the constraints on $F(x, y)$, $B(x, y)$ are based on the assumption that $F(x, y)$ and $B(x, y)$ are smooth functions.

A couple of matting models have been proposed such as the Poisson matting, Bayes matting, RW matting, closed-form matting, robust matting, etc. Sometimes, user input is required to identify foreground, background, and transitional regions, which are referred to as the trimap [13]. In the current context, a trimap can be generated automatically from the segmentation step by extracting the object boundary bound [103]. The Soft Scissor (SS) [111] offers a real-time interactive matting tool, and it is implemented as Digital Film Tools Power Mask (http://www.digitalfilmtools. com/powermask/) with a user input similiar to that of edge-based Intelligent Scissors [41].

Another efficient boundary refinement technique is to apply an active contour method with an additional constraint [112] that is used to determine a global optimal object boundary. It aims to locate the most likely object boundary by considering both the boundary and the regional information. For segmentation methods with a hard segmentation result, a probability map based on the GMM models of the foreground and background colors is first constructed. Then, the constrained active contour technique is adopted to find the optimal boundary location. As shown in [112], this technique is effective in improving randow-walk methods [6] and the geodesic segmentation method [13] by generating an improved hard segmentation result.

To conclude, boundary editing and isolated region merging offer two post-processing solutions in image segmentation with explicit control. The alpha matting technique is particularly efficient in handling hairy and furry complex boundaries.

References

1. Boykov Y, Jolly M (2001) Interactive graph cuts for optimal boundary & region segmentation of objects in nd images. In: Proceeding of 8th IEEE international conference on computer vision, ICCV 2001, IEEE, vol 1, pp 105–112

2. Greig D, Porteous B, Seheult A (1989) Exact maximum a posteriori estimation for binary images. J Roy Stat Soc 51(2):271–279
3. McGuinness K, O'Connor N (2010) A comparative evaluation of interactive segmentation algorithms. Pattern Recogn 43(2):434–444
4. Rother C, Kolmogorov V, Blake A (2004) "Grabcut": interactive foreground extraction using iterated graph cuts. ACM Trans Graph 23(3):309–314
5. Li Y, Sun J, Tang CK, Shum HY (2004) Lazy snapping. ACM Trans Graph 23(3):303–308
6. Boykov Y, Funka-Lea G (2006) Graph cuts and efficient n-d image segmentation. Int J Comput Vision 70(2):109–131
7. Boykov Y, Veksler O (2006) Graph cuts in vision and graphics: theories and applications. Handbook of mathematical models in computer vision, pp. 79–96
8. Boykov Y, Kolmogorov V (2004) An experimental comparison of min-cut/max-flow algorithms for energy minimization in vision. IEEE Trans Pattern Anal Mach Intell 26(9):1124–1137
9. Ford LR, Fulkerson DR (1962) Flows in networks. Princeton University Press, Princeton
10. Wu Z, Leahy R (1993) An optimal graph theoretic approach to data clustering: theory and its application to image segmentation. IEEE Trans Pattern Anal Mach Intell 15(11):1101–1113
11. Vincent L, Soille P (1991) Watersheds in digital spaces: an efficient algorithm based on immersion simulations. IEEE Trans Pattern Anal Mach Intell 13(6):583–598
12. Price B, Morse B, Cohen S (2010) Geodesic graph cut for interactive image segmentation. In: 2010 IEEE conference on computer vision and pattern recognition, IEEE, pp 3161–3168
13. Bai X, Sapiro G (2007) A geodesic framework for fast interactive image and video segmentation and matting. In: IEEE 11th international conference on computer vision, ICCV 2007, IEEE, pp 1–8
14. Criminisi A, Sharp T, Blake A (2008) Geos: geodesic image segmentation. Comput Vis ECCV 2008:99–112
15. Freedman D, Zhang T (2005) Interactive graph cut based segmentation with shape priors. In: IEEE computer society conference on computer vision and pattern recognition, IEEE, vol 1, pp 755–762
16. Veksler O (2008) Star shape prior for graph-cut image segmentation. Comput Vis ECCV 2008:454–467
17. Gulshan V, Rother C, Criminisi A, Blake A, Zisserman A (2010) Geodesic star convexity for interactive image segmentation. In: 2010 IEEE conference on computer vision and pattern recognition(CVPR), IEEE, pp 3129–3136
18. Vicente S, Kolmogorov V, Rother C (2008) Graph cut based image segmentation with connectivity priors. In: IEEE Conference on Computer vision and pattern recognition, CVPR 2008, IEEE, pp 1–8
19. Wang J, Bhat P, Colburn R, Agrawala M, Cohen M (2005) Interactive video cutout. In: ACM transactions on graphics, ACM, pp 585–594
20. Lombaert H, Sun Y, Grady L, Xu C (2005) A multilevel banded graph cuts method for fast image segmentation. In: 10th IEEE international conference on computer vision, ICCV 2005, IEEE, vol 1, pp 259–265
21. Pratt W (2007) Digital image processing: PIKS scientific inside. Wiley-Interscience Publication, Hoboken
22. Lakshmi S, Sankaranarayanan D (2010) A study of edge detection techniques for segmentation computing approaches. Int J Comput Appl IJCA 1:7–10
23. Canny J (1986) A computational approach to edge detection. IEEE Trans Pattern Anal Mach Intell 6:679–698
24. Bao P, Zhang L, Wu X (2005) Canny edge detection enhancement by scale multiplication. IEEE Trans Pattern Anal Mach Intell 27(9):1485–1490
25. Demigny D (2002) On optimal linear filtering for edge detection. IEEE Trans Image Process 11(7):728–737
26. Petrou M, Kittler J (1991) Optimal edge detectors for ramp edges. IEEE Trans Pattern Anal Mach Intell 13(5):483–491

27. Torre V, Poggio T (1986) On edge detection. IEEE Trans Pattern Anal Mach Intell 2:147–163
28. Marr D, Hildreth E (1980) Theory of edge detection. Proc Roy Soc Lond Ser B Biol Sci 207(1167):187–217
29. Haralick R (1984) Digital step edges from zero crossing of second directional derivatives. IEEE Trans Pattern Anal Mach Intell 1:58–68
30. Gauch J, Hsia C (1992) Comparison of three-color image segmentation algorithms in four color spaces. In: Applications in optical science and engineering. International Society for Optics and Photonics, Bellingham, pp 1168–1181
31. Koschan A, Abidi M (2005) Detection and classification of edges in color images. IEEE Signal Process Mag 22(1):64–73
32. Mortensen E, Morse B, Barrett W, Udupa J (1992) Adaptive boundary detection using 'live-wire' two-dimensional dynamic programming. In: Proceedings of computers in cardiology, IEEE, pp 635–638
33. Udupa J, Samarasekera S, Barrett W (1992) Boundary detection via dynamic programming. In: Proceedings of SPIE-the international society for optical engineering, pp 33–33
34. Barrett W, Mortensen E (1996) Fast, accurate, and reproducible live-wire boundary extraction. In: Visualization in biomedical computing. Springer, New York, pp 183–192
35. Barrett W, Mortensen E (1997) Interactive live-wire boundary extraction. Med Image Anal 1(4):331–341
36. Falcão A, Udupa J, Miyazawa F (2000) An ultra-fast user-steered image segmentation paradigm: live wire on the fly. IEEE Trans Med Imaging 19(1):55–62
37. Falcão A, Udupa J, Samarasekera S, Hirsch B (1996) User-steered image boundary segmentation. In: Proceedings of SPIE on medical imaging, vol 2710, pp 278–288
38. Falcão A, Udupa J, Samarasekera S, Sharma S, Hirsch B, Lotufo R (1998) User-steered image segmentation paradigms: live wire and live lane. Graph Models Image Process 60(4):233–260
39. Kang H, Shin S (2002) Enhanced lane: interactive image segmentation by incremental path map construction. Graph Models 64(5):282–303
40. Mortensen E, Barrett W (1998) Interactive segmentation with intelligent scissors. Graph Models Image Process 60(5):349–384
41. Mortensen EN, Barrett WA (1995) Intelligent scissors for image composition. In: Proceedings of the 22nd annual conference on computer graphics and interactive techniques, SIGGRAPH '95. ACM, New York, pp 191–198
42. Gimp G (2008) Image manipulation program. User manual, edge-detect filters, sobel. The GIMP Documentation Team
43. Dijkstra E (1959) A note on two problems in connexion with graphs. Numer math 1(1):269–271
44. Mortensen E, Barrett W (1999) Toboggan-based intelligent scissors with a four-parameter edge model. In: IEEE computer society conference on computer vision and pattern recognition, IEEE, vol 2
45. Blake A, Isard M et al (1998) Active contours, vol 1. Springer, London
46. McInerney T, Terzopoulos D (1996) Deformable models in medical image analysis: a survey. In: Proceedings of the workshop on mathematical methods in biomedical image analysis, IEEE, pp 171–180
47. Singh A, Terzopoulos D, Goldgof D (1998) Deformable models in medical image analysis. IEEE Computer Society Press, New York
48. Kass M, Witkin A, Terzopoulos D (1988) Snakes: active contour models. Int J Comput Vis 1(4):321–331
49. Papari G, Petkov N (2011) Edge and line oriented contour detection: state of the art. Image Vis Comput 29(2):79–103
50. Brigger P, Hoeg J, Unser M (2000) B-spline snakes: a flexible tool for parametric contour detection. IEEE Trans Image Process 9(9):1484–1496
51. Cohen L, Cohen I (1993) Finite-element methods for active contour models and balloons for 2-d and 3-d images. IEEE Trans Pattern Anal Mach Intell 15(11):1131–1147

52. Gunn S, Nixon M (1995) Improving snake performance via a dual active contour. In: Computer analysis of images and patterns. Springer, Berlin, pp 600–605
53. Gunn S, Nixon M (1997) A robust snake implementation; a dual active contour. IEEE Trans Pattern Anal Mach Intell 19(1):63–68
54. Leymarie F, Levine M (1993) Tracking deformable objects in the plane using an active contour model. IEEE Trans Pattern Anal Mach Intell 15(6):617–634
55. Li B, Acton S (2007) Active contour external force using vector field convolution for image segmentation. IEEE Trans Image Process 16(8):2096–2106
56. Mishra A, Fieguth P, Clausi D (2011) Decoupled active contour (dac) for boundary detection. IEEE Trans Pattern Anal Mach Intell 33(2):310–324
57. Shih F, Zhang K (2007) Locating object contours in complex background using improved snakes. Comput Vis Image Underst 105(2):93–98
58. Sundaramoorthi G, Yezzi A, Mennucci A (2008) Coarse-to-fine segmentation and tracking using sobolev active contours. IEEE Trans Pattern Anal Mach Intell 30(5):851–864
59. Wong Y, Yuen P, Tong C (1998) Segmented snake for contour detection. Pattern Recogn 31(11):1669–1679
60. Xu C, Prince J (1998) Snakes, shapes, and gradient vector flow. IEEE Trans Image Process 7(3):359–369
61. Bresson X, Esedoglu S, Vandergheynst P, Thiran J, Osher S (2007) Fast global minimization of the active contour/snake model. J Math Imaging Vis 28(2):151–167
62. Caselles V, Catté F, Coll T, Dibos F (1993) A geometric model for active contours in image processing. Numer Math 66(1):1–31
63. Chan T, Vese L (2001) Active contours without edges. IEEE Trans Image Process 10(2):266–277
64. Malladi R, Sethian J, Vemuri B (1995) Shape modeling with front propagation: a level set approach. IEEE Transa Pattern Anal Mach Intell 17(2):158–175
65. Zhu S, Yuille A (1996) Region competition: Unifying snakes, region growing, and bayes/mdl for multiband image segmentation. IEEE Trans Pattern Anal Mach Intell 18(9):884–900
66. Jalba A, Wilkinson M, Roerdink J (2004) Cpm: a deformable model for shape recovery and segmentation based on charged particles. IEEE Trans Pattern Anal Mach Intell 26(10):1320–1335
67. Xie X, Mirmehdi M (2008) Mac: magnetostatic active contour model. IEEE Trans Pattern Anal Mach Intell 30(4):632–646
68. Liang J, McInerney T, Terzopoulos D et al (2006) United snakes. Med Image Anal 10(2):215–233
69. Zienkiewicz O, Taylor R, Zhu J (2005) The finite element method: its basis and fundamentals, vol 1. Butterworth-Heinemann, Oxford
70. Ronfard R (1994) Region-based strategies for active contour models. Int J Comput Vis 13(2):229–251
71. Paragios N, Deriche R (2002) Geodesic active regions and level set methods for supervised texture segmentation. Int J Comput Vis 46(3):223–247
72. Leventon M, Grimson W, Faugeras O (2000) Statistical shape influence in geodesic active contours. In: Proceedings of IEEE conference on computer vision and pattern recognition, IEEE, vol 1, pp 316–323
73. Nguyen TNA, Cai J, Zhang J, Zheng J (2012) Robust interactive image segmentation using convex active contours. IEEE Transactions on Image Processing 21(8):3734–3743
74. Koller D, Friedman N (2009) Probabilistic graphical models: principles and techniques. MIT press, Cambridge
75. TT II C, EE C, Image segmentation and techniques: a review
76. Grady L (2006) Random walks for image segmentation. IEEE Trans Pattern Anal Mach Intell 28(11):1768–1783
77. Grady L, Sinop A (2008) Fast approximate random walker segmentation using eigenvector precomputation. In: IEEE conference on computer vision and pattern recognition, CVPR 2008, IEEE, pp 1–8

78. Kim T, Lee K, Lee S (2008) Generative image segmentation using random walks with restart. Comput Vis ECCV 2008:264–275
79. Courant R, Hilbert D (1944) Methods of mathematical physics. Interscience Publishers Inc., New York
80. Dodziuk J (1984) Difference equations, isoperimetric inequality and transience of certain random walks. Trans Am Math Soc 284:787–794
81. Yang W, Cai J, Zheng J, Luo J (2010) User-friendly interactive image segmentation through unified combinatorial user inputs. IEEE Trans Image Process 19(9):2470–2479
82. Pan J, Yang H, Faloutsos C, Duygulu P (2004) Automatic multimedia cross-modal correlation discovery. In: Proceedings of the 10th ACM SIGKDD international conference on knowledge discovery and data mining, ACM, pp 653–658
83. Tong H, Faloutsos C, Pan J (2008) Random walk with restart: fast solutions and applications. Knowl Inf Syst 14(3):327–346
84. Jeh G, Widom J (2003) Scaling personalized web search. In: Proceedings of the 12th international conference on World Wide Web, ACM, pp 271–279
85. Kamvar S, Haveliwala T, Manning C, Golub G (2003) Exploiting the block structure of the web for computing pagerank. Stanford University Technical Report
86. Fogaras D, Rácz B (2004) Towards scaling fully personalized pagerank. In: Algorithms and Models for the Web-Graph, pp 105–117
87. Beucher S et al (1992) The watershed transformation applied to image segmentation. Scanning Microsc Suppl 6:299–314
88. Adams R, Bischof L (1994) Seeded region growing. IEEE Trans Pattern Anal Mach Intell 16(6):641–647
89. Horowitz SL, Pavlidis T (1974) Picture segmentation by a directed split-and-merge procedure. In: Proceedings of the 2nd international joint conference on pattern recognition, vol 424, p 433
90. Raja D, Khadir A, Ahamed D (2009) Moving toward region-based image segmentation techniques: a study. J Theor Appl Inf Technol 81–87
91. Haris K, Efstratiadis S, Maglaveras N, Katsaggelos A (1998) Hybrid image segmentation using watersheds and fast region merging. IEEE Trans Image Process 7(12):1684–1699
92. Mohana Rao K, Dempster A (2002) Modification on distance transform to avoid over-segmentation and under-segmentation. In: Video/Image processing and multimedia communications 4th EURASIP-IEEE region 8 international symposium on VIPromCom, IEEE, pp 295–301
93. Edge detection and image segmentation (edison) system (2009). http://coewww.rutgers.edu/riul/research/code.html
94. Mehnert A, Jackway P (1997) An improved seeded region growing algorithm. Pattern Recogn Lett 18(10):1065–1071
95. Fan J, Zeng G, Body M, Hacid M (2005) Seeded region growing: an extensive and comparative study. Pattern Recogn Lett 26(8):1139–1156
96. Vezhnevets V, Konouchine V (2005) Growcut: interactive multi-label nd image segmentation by cellular automata. In: Proceedings of graphicon, pp 150–156
97. Von Neumann J, Burks A (1966) Theory of self-reproducing automata. University of Illinois Press, Urbana
98. Das D (2012) A survey on cellular automata and its applications. In: Global trends in computing and communication systems, pp 753–762
99. Kari J (2005) Theory of cellular automata: a survey. Theor Comput Sci 334(1):3–33
100. Kauffmann C, Piché N (2010) Seeded and medical image segmentation by cellular automaton on GPU. Int J Comput Assist Radiol Surg 5(3):251–262
101. Grady L, Funka-Lea G (2004) Multi-label image segmentation for medical applications based on graph-theoretic electrical potentials. In: Computer vision and mathematical methods in medical and biomedical image, analysis, pp 230–245
102. Ning J, Zhang L, Zhang D, Wu C (2010) Interactive image segmentation by maximal similarity based region merging. Pattern Recogn 43(2):445–456

103. Noma A, Graciano A, Consularo L, Bloch I (2012) Interactive image segmentation by match-
     ing attributed relational graphs. Pattern Recogn 45(3):1159–1179
104. Consularo L, Cesar R, Bloch I (2007) Structural image segmentation with interactive model
     generation. In: IEEE international conference on image processing, vol 6, IEEE, pp VI-45
105. Bunke H (2000) Recent developments in graph matching. In: 15th international conference
     on pattern recognition, vol 2, IEEE, pp 117–124
106. Noma A, Pardo A (2011) Structural matching of 2d electrophoresis gels using deformed
     graphs. Pattern Recogn Lett 32(1):3–11
107. Lin H, Tai Y, Brown M (2011) Motion regularization for matting motion blurred objects.
     IEEE Trans Pattern Anal Mach Intell 33(11):2329–2336
108. Porter T, Duff T (1984) Compositing digital images. ACM Siggraph Comput Graph
     18(3):253–259
109. Wang J, Cohen MF (2005) An iterative optimization approach for unified image segmentation
     and matting. In: 10th IEEE international conference on computer vision, ICCV 2005, vol 2,
     IEEE, pp 936–943
110. Wang J, Cohen MF (2008) Image and video matting, vol 3. Now Publishers, Hanover
111. Wang J, Agrawala M, Cohen MF (2007) Soft scissors: an interactive tool for realtime high
     quality matting. In: ACM Transactions on Graphics (TOG), vol 26, p 9, ACM
112. Anh NTN, Cai J, Zhang J, Zheng J (2012) Constrained active contours for boundary refinement
     in interactive image segmentation. In: 2012 IEEE international symposium on circuits and
     systems (ISCAS), IEEE, pp 870–873

Chapter 4
Performance Evaluation

Keywords Interactive graph-cut · Random walks with restart · Convex active contour · Maximal similarity-based region merging · Matching attributed relational graph

Interactive segmentation methods are developed to solve the image segmentation problem in real-world applications. It is desirable that, with user interactions, segmentation techniques can segment out arbitrary objects of interest accurately. The procedure should be intelligent and easily controllable by users. Nevertheless, there is a gap between this goal and what today's algorithms can offer. In this chapter, we evaluate the performance of several state-of-the-art interactive segmentation methods with a set of "challenging" images.

The test image set includes those that have dull colors, low contrast, elongated objects, objects with weak boundaries, cluttered background, and a strong noise component. The methods under evaluation include:

- the interactive graph-cut (IGC) method [1],
- the random-walk with restart (RWR) method [2],
- the convex active contour (CAC) method [3],
- the maximal similarity-based region merging (MSRM) method [4],
- the matching attributed relational graph (MARG) method [5].

These methods were elaborated in Chap. 3.

4.1 Similarity Measures

One common task in both pixel-based and region-based segmentation methods is the measure of similarities between adjacent pixels (or super-pixels) so that one can give them the same label or different labels. Similarity measures can be classified into two categories:

- Appearance similarity

  Luminance and color intensities can be used to measure the appearance similarity. As discussed in Chap. 3, these features are used to compute the similarities between a node and seed nodes. For example, in the IGC method [1], image intensities are modeled as the luminance histogram, which is less sensitive to color variations of objects. The GrabCut method [6] uses a color GMM model. It performs well when objects have colors that are very different from that of the background. However, when the object and background have similar colors, GrabCut may fail to extract objects properly with a simple rectangular user markup as illustrated in Fig. 4.1. Further user's scribbles are required to generate acceptable results.

- Structure similarity [1, 5, 7]

  The definition of structure similarity affects the segmentation performance. For example, in some cases where the object has colors that are distinctive from those of the background, the graph-cut method using the Euclidean distance suffers from the short path problem, yielding incomplete object segmentation as shown

Fig. 4.1 A *rectangular markup* is not sufficient for the GrabCut method to extract an object when it and its background have similar colors. **a** Original image, where *red* and *blue* scribbles indicate the object region and background, respectively. **b** Segmentation result by IGC. **c** Original image with a *red rectangular mark* indicating the object. **d** Segmentation result by GrabCut

Fig. 4.2 Illustration of the short path problem of the IGC method. **a** Original image with user scribbles, where *white* and *red* indicate the object and background, respectively. **b** Segmentation result by IGC, with missing cow legs. **c** Segmentation result by GGC, where cow legs are segmented correctly

in Fig. 4.2. In this example, the geodesic graph-cut (GGC) [7] performs better than IGC based on the same scribbles from users. This is because the GGC method takes the local path cost into account in the distance measure. The structure similarity and constraints are important to maintain the completeness of object segmentation. This observation was discussed before, e.g., [5, 8, 9].

We see from these two examples that a proper similarity definition will have a major impact on segmentation results. To achieve better results, we may allow different similarity definitions based on the characteristics of input images. Besides, the parameters of each method can be fine-tuned to yield better results.

4.2 Evaluation on Challenging Images

In this section, we compare the performance of several representative methods discussed in Chap. 3 on a set of challenging images. Our comparative results serve as an extended evaluation of prior work in [5, 10–13]. Note that better results can be achieved by adding more user labels. For fair comparison, we compare the performances of different methods with the same user input here.

4.2.1 Images with Similar Foreground and Background Colors

The first test image in Fig. 4.3 contains a ceramic model with two main colors, blue and yellow, placed on a table that has a similar yellow color. The blue parts of the ceramic model can be easily distinguished from the image. However, the yellow parts have colors similar to the background, yielding weak boundaries. Also, the overall contrast of the test image is low. It is therefore challenging to segment the complete object out with limited user inputs as depicted in Fig. 4.3a.

We have the following observations on the results given in Fig. 4.3. First, RWR and MARG fail to segment the object out. RWR is sensitive to the locations of user

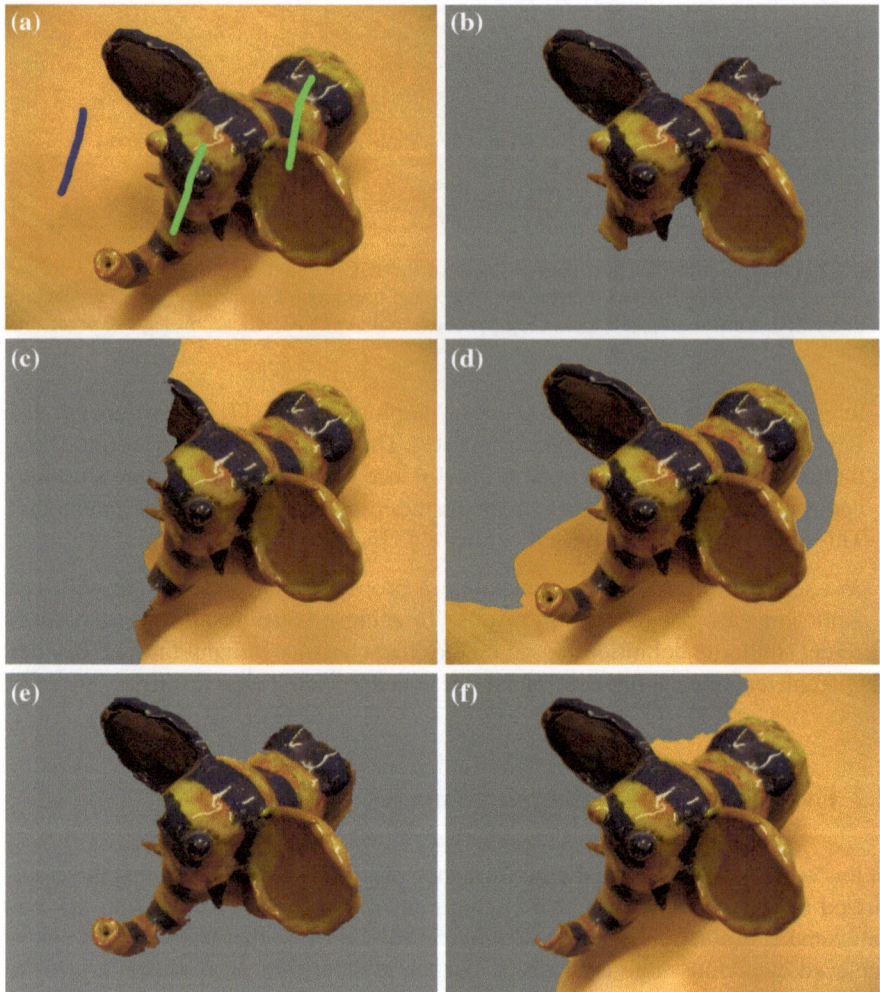

Fig. 4.3 Performance comparison on a test image, where the foreground object and background share similar colors. **a** Original image with user inputs, in which *green* and *blue* scribbles indicate the object and background, respectively. **b** Segmentation result by IGC [1]. **c** Segmentation result by RWR [2]. **d** Segmentation result by CAC [3]. **e** Segmentation result by MSRM [11]. **f** Segmentation result by MARG [5]

scribbles. Without any label on the right side of background, this region is falsely declared to be the foreground. Similarly, MARG is also sensitive to scribble locations, since it considers the graphical structure of user labels. IGC, which uses a luminance histogram model, can segment out the main part of the background. However, some parts, even blue parts, are missing, since those parts have a luminance level similar to that of background. CAC attempts to find an accurate object contour with boundary refinement, yet its result depends on the primary segmentation. MSRM and MARG use superpixels. Although the foreground and background have similar colors, a good pre-segmentation procedure can cluster foreground and background pixels into different superpixels. In this test, MSRM produces the best result since it merges superpixels gradually and updates color histogram models at each iteration. However, some weak boundaries due to the bottom shadow are still incorrectly segmented. Since all segmentation methods depend on color and edge features, they demand more user scribbles to segment this test image correctly.

4.2.2 Images with Complex Contents

Besides similar foreground and background colors, the variety and complexity of foreground and background contents pose a challenge to segmentation tools. Note that, even if the foreground object and background may have different colors, seed pixels in the foreground object and background will have a wide range of feature values due to complex contents.

The test image in Fig. 4.4 contains a banana on a textured table. Both the banana and the table have complex but little overlapping colors. IGC and RWR fail to segment the banana in the image. In this test, IGC is not able to use the color information effectively and, thus, the spatial distance dominates the similarity measure, yielding an incorrect result. Since roses in background are not well connected, RWR cannot make all roses reachable with high probabilities from background seeds. Clearly, more user scribbles are needed for RWR. In contrast, CAC, MSRM and MARG extract the banana more accurately. CAC provides a more accurate boundary than MSRM and MARG by employing the convex active contour.

Figure 4.5 gives another complex image, where the object has a net structure. The object is connected yet with holes. It is demanding to ask a user to label all background holes, since there are too many isolated parts. Besides, some background parts are blended with the foreground net. With limited user inputs as shown in Fig. 4.4a, CAC can generate an acceptable result, while others fail to segment some obvious background parts from the net. The results of MSRM and MARG are influenced by their pre-segmented superpixels. They both tend to merge an unlabeled hole into its neighbor, which is not desirable in extracting the net structure.

4.2.3 Images with Multiple Objects

In this book, our main focus is on extracting a single object from the background. To segment multiple objects separately, the label propagation technique can be extended to accept more than two object labels. To give an example, we attempt to extract two birds as a single foreground object in Fig. 4.6.

The background is smooth in the "Two birds" test image, and it should be relatively easy to segment the two birds out. However, with simple user inputs as shown in

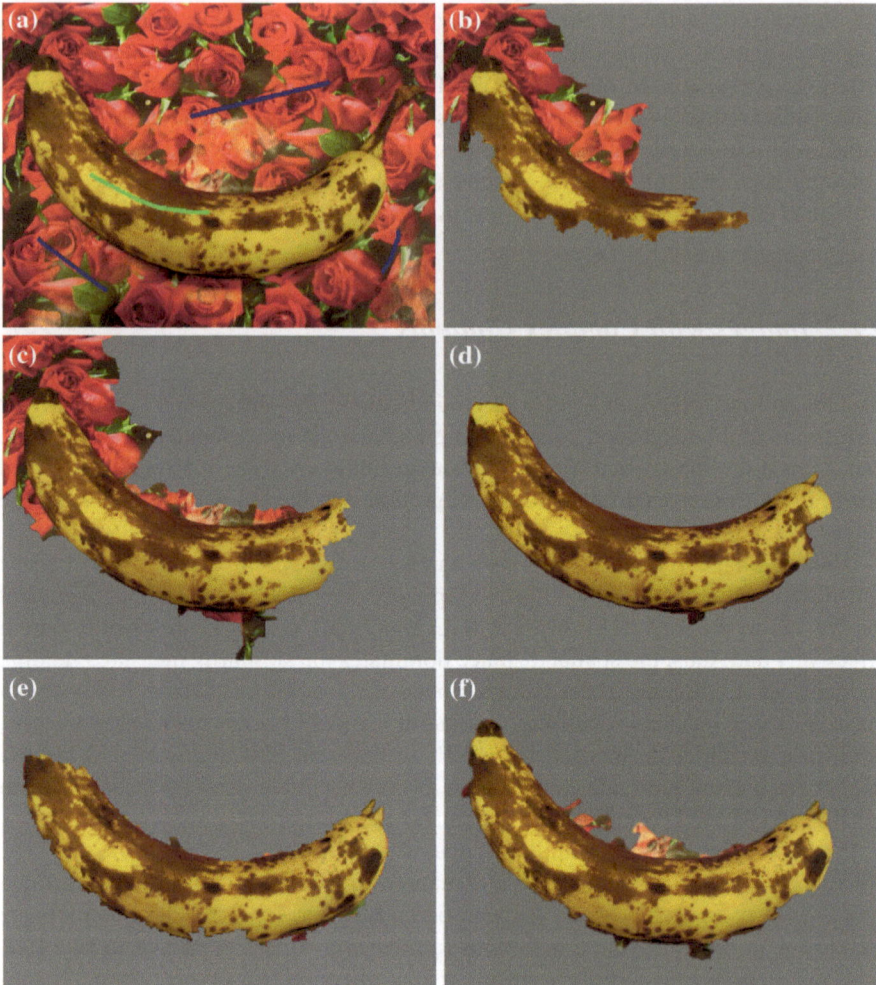

Fig. 4.4 Performance comparison on the segmentation of the "Banana" image with cluttered background. **a** Original image with user scribbles. **b** Segmentation result by IGC [1]. **c** Segmentation result by RWR [2]. **d** Segmentation result by CAC [3]. **e** Segmentation result by MSRM [5]. **f** Segmentation result by MARG [4]

Fig. 4.6a, IGC and RWR cannot extract the birds completely. IGC misses the wings and mouths of the birds. RWR fails to identify the background, although it has a flat and unique color. These results can be improved by adding more user scribbles. The other three methods can segment the main parts of the birds out, but miss the elongated mouths. To make the extracted object complete is an important issue. This problem can be improved by GGC with the shape and connectivity priors [7, 14, 15].

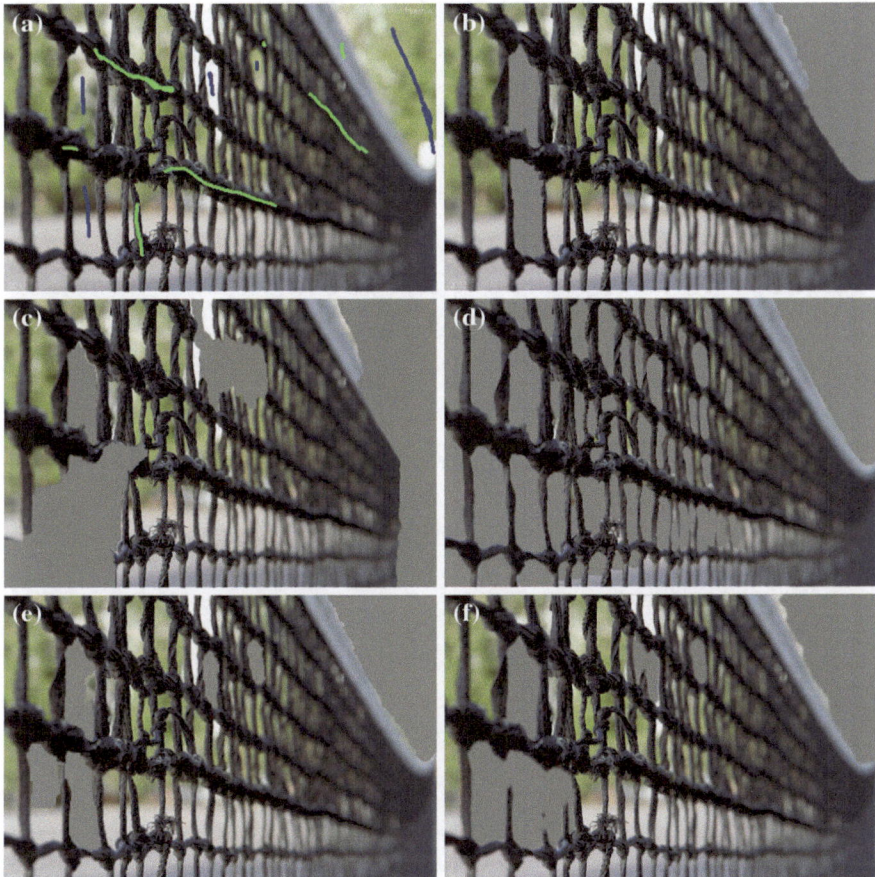

Fig. 4.5 Performance comparison on the segmentation of the "Net" image with many isolated background parts. **a** Original image with user scribbles. **b** Segmentation result by IGC [1]. **c** Segmentation result by RWR [2]. **d** Segmentation result by CAC [3]. **e** Segmentation result by MSRM [11]. **f** Segmentation result by MARG [5]

4.2.4 Images with Noise

Digital images contain noise such as acquisition noise and transmission noise. The advancement of acquisition sensors and denoising techniques can reduce noise greatly [16]. However, we still encounter noisy images in practical applications. In this test, we investigate the robustness of segmentation methods applied to images with mixed Gaussian and salt-and-pepper noise.

We see from Fig. 4.7 that the superpixel-based approaches, MSRM and MARG, can produce similar results on images with and without noise. Also, CAC can locate similar object contours. These results indicate that MSRM, MARG, and CAC are more robust to noise than IGC and RWR.

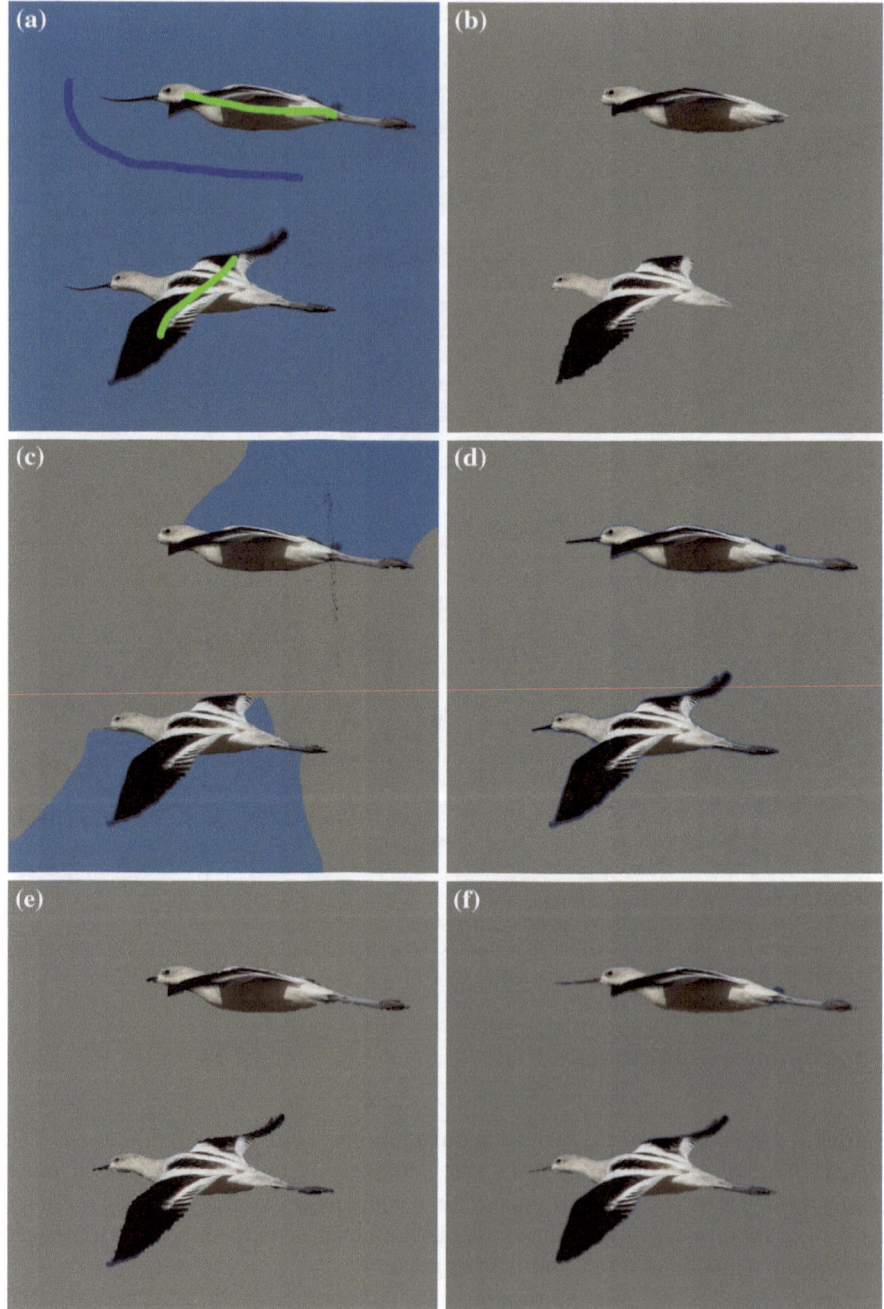

Fig. 4.6 Performance comparison on the segmentation of "Two birds" as one foreground object. **a** Original image with user scribbles. **b** Segmentation result by IGC [1]. **c** Segmentation result by RWR [2]. **d** Segmentation result by CAC [3]. **e** Segmentation result by MSRM [11]. **f** Segmentation result by MARG [5]

Fig. 4.7 An example of segmenting original and noisy images. **a** Original image with user scribbles. **b** Segmentation of original image by IGC. **c** Noisy image with user scribbles. **d** Segmentation of noisy image by IGC. **e** Segmentation of original image by RWR. **f** Segmentation of noisy image by RWR. **g** Segmentation of original image by CAC. **h** Segmentation of noisy image by CAC. **i** Segmentation of original image by MSRM. **j** Segmentation of noisy image by MSRM. **k** Segmentation of original image by MARG. **l** Segmentation of noisy image by MARG

Besides noisy images, images of fog and rain have many small and translucent particles. They can be treated as noise when we intend to segment objects out. In Fig. 4.8, we show an image that is covered with a spray of various degree. The segmentation target is the little boy in the middle. The object is blended with heterogeneous spray noise and the image contrast is low due to the translucent noise. In this test, IGC and CAC perform better than others. RWR fails to identify object boundaries. MSRM and MARG, which are based on pre-segmentation, cannot segment the object correctly when the object regions falsely merged with background in the pre-processing step.

4.3 Discussion

Generally speaking, the performance of interactive segmentation methods can be evaluated in terms of regional accuracy, boundary accuracy, the running speed, the user interaction requirement, and the memory requirement [5, 10–13].

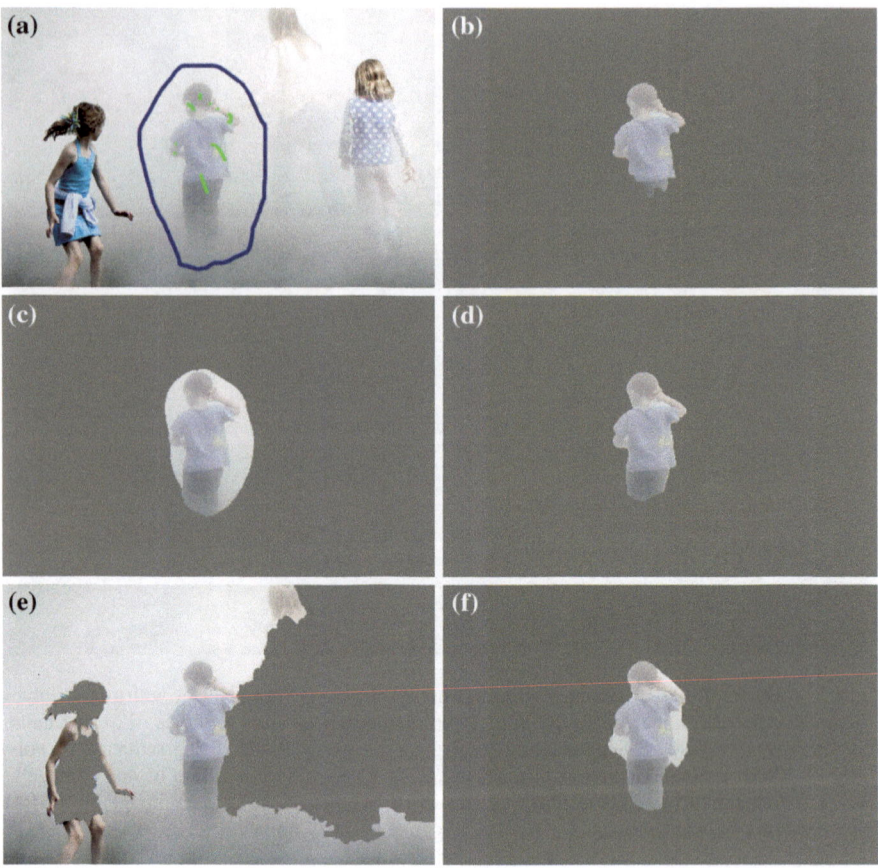

Fig. 4.8 Segmentation of a hazy image. **a** Original image with user scribbles. **b** Segmentation result by IGC [1]. **c** Segmentation result by RWR [2]. **d** Segmentation result by CAC [3]. **e** Segmentation result by MSRM [11]. **f** Segmentation result by MARG [5]

In this chapter, we focused more on the issues of accuracy and robustness, and evaluated several popular methods on a couple of challenging images under the same user input. Specifically, we have the following observations:

- The IGC method has the short path problem and may fail on complicated image contents.
- The RWR method is sensitive to user inputs and noise components.
- The CAC method outperforms others on the net-structured image.
- The MSRM and the MARG methods provide similar performances in most cases while the MARG method is more sensitive to scribble locations.

References

1. Boykov Y, Funka-Lea G (2006) Graph cuts and efficient n-d image. In: Paragios N, Chen Y, Faugeras O (eds) Int J Comput Vis 70(2):109–131
2. Kim T, Lee K, Lee S (2008) Generative image segmentation using random walks with restart. In: Computer vision ECCV 2008, pp 264–275
3. Nguyen TNA, Cai J, Zhang J, Zheng J (2012) Robust interactive image segmentation using convex active contours. IEEE Trans Image Process 21(8):3734–3743
4. Ning J, Zhang L, Zhang D, Wu C (2010) Interactive image segmentation by maximal similarity based region merging. Pattern Recogn 43(2):445–456
5. Noma A, Graciano A, Consularo L, Bloch I (2012) Interactive image segmentation by matching attributed relational graphs. Pattern Recogn 45(3):1159–1179
6. Rother C, Kolmogorov V, Blake A (2004) "GrabCut": interactive foreground extraction using iterated graph cuts. ACM Trans Graph 23(3):309–314
7. Price B, Morse B, Cohen S (2010) Geodesic graph cut for interactive image segmentation. In: 2010 IEEE conference on computer vision and pattern recognition. IEEE, pp 3161–3168
8. Gulshan V, Rother C, Criminisi A, Blake A, Zisserman A (2010) Geodesic star convexity for interactive image segmentation. In: 2010 IEEE conference on computer vision and pattern recognition (CVPR). IEEE, pp 3129–3136
9. Veksler O (2008) Star shape prior for graph-cut image segmentation. In: Computer vision ECCV 2008, pp 454–467
10. Grady L, Sun Y, Williams J (2006) Three interactive graph-based segmentation methods applied to cardiovascular imaging. In: Handbook of mathematical models in computer vision, pp 453–469
11. McGuinness K, O'Connor N (2010) A comparative evaluation of interactive segmentation algorithms. Pattern Recogn 43(2):434–444
12. Moschidis E, Graham J (2009) Simulation of user interaction for performance evaluation of interactive image segmentation methods. In: Proceedings of the 13th medical image understanding and analysis conference, pp 209–213
13. Yang W, Cai J, Zheng J, Luo J (2010) User-friendly interactive image segmentation through unified combinatorial user inputs. IEEE Trans Image Process 19(9):2470–2479
14. Freedman D, Zhang T (2005) Interactive graph cut based segmentation with shape priors. In: IEEE computer society conference on computer vision and pattern recognition, vol 1. IEEE, pp 755–762
15. Vicente S, Kolmogorov V, Rother C (2008) Graph cut based image segmentation with connectivity priors. In: IEEE conference on computer vision and pattern recognition CVPR 2008. IEEE, pp 1–8
16. Gonzalez RC, Woods RE (2007) Digital image processing, 3rd edn. Prentice Hall, New Jersey

Chapter 5
Conclusion and Future Work

Keywords Segmentation accuracy · Robustness of user interaction · Dynamic interaction · 3D image segmentation

Interactive segmentation techniques have attracted a wide range of interest and applications. Many researchers have worked on this topic to improve the efficiency, robustness, speed, and user-friendliness of interactive segmentation. Image features such as colors, edges, and locations are essential for computers to recognize and extract objects. By employing various principles such as graph-cut, random-walk, or region merging/splitting, interactive segmentation methods attempt to balance two constraints: regional-homogenuity and boundary-inhomogenuity. These two constraints are expressed as a cost function in most segmentation methods which is then optimized locally and/or globally. Satisfactory results can be obtained by incorporating a sufficient amount of user interactions.

For these methods to be applicable in real-world applications, more future research is needed along the following three major directions.

- Accuracy of segmentation
  We tested several state-of-the-art methods and showed that each of them has its own limitations. How to improve segmentation accuracy for a wide range of images is an ongoing topic. The region-level accuracy is required to enforce the completeness of segmentation results while the pixel-level accuracy is important in locating accurate object boundaries. Soft segmentation with the alpha matte may be needed for complex boundaries.
- Robustness of user interaction
  Although interactive segmentation methods allow users to label foreground objects and background to facilitate the segmentation process, many segmentation results are sensitive to the location of initial labels. A good set of initial labels that achieves the desired goal is highly dependent on image content. It tends to take a beginner

a long time to learn a proper way to do the labeling. This shortcoming will hinder the applicability of interactive segmentation methods.

- Two-way dynamic interaction
  It is desirable that the computer analyzes the image content and, then, guides users to provide their input to facilitate the segmentation task. After the first round of segmentation, the user should only highlight the region that is not satisfactory and the computer can react with further refinement. How to make the two-way interaction more effective so as to reduce the number of interactive segmentation rounds is an important topic.

Finally, interactive segmentation tools should be generalized to 3D images and video. So far, there is little work along this direction. It is not a trivial task to design friendly tools for visualization and user interaction on these high-dimensional data.